Julia Lieb

Counting Polynomial Matrices over Finite Fields

Julia Lieb

Counting Polynomial Matrices over Finite Fields

Matrices with Certain Primeness Properties and Applications to Linear Systems and Coding Theory

Würzburg
University Press

Dissertation, Julius-Maximilians-Universität Würzburg
Fakultät für Mathematik und Informatik, 2017
Gutachter: Prof. Dr. Peter Müller, Prof. Dr. Joachim Rosenthal

Impressum

Julius-Maximilians-Universität Würzburg
Würzburg University Press
Universitätsbibliothek Würzburg
Am Hubland
D-97074 Würzburg
www.wup.uni-wuerzburg.de

© 2017 Würzburg University Press
Print on Demand

Coverdesign: Jule Petzold

ISBN 978-3-95826-064-1 (print)
ISBN 978-3-95826-065-8 (online)
URN urn:nbn:de:bvb:20-opus-151303

To Monika and Rudolf.

Acknowledgments

I would like to express my sincere gratitude to Prof. Uwe Helmke for his supervision, fruitful discussions and the opportunity to work under excellent research conditions at the Chair for "Dynamical Systems and Control Theory".

I want to thank Prof. Müller and Prof. Rosenthal for their immediate help and agreement to continue the supervision of this thesis as well as for supporting me in the period of finishing this dissertation.

Especially, I would like to thank Joachim Rosenthal for giving me the opportunity to visit his research group at the University of Zurich for the fall term 2016 and for his great hospitality.

I thank all my colleagues from the Chair of "Dynamical Systems and Control Theory" at the University of Würzburg and from the Chair of "Applied Algebra" at the University of Zurich.

This thesis would not have been written without the help and support of my friends and family. I would like to thank them for everything.

Julia Lieb

Contents

Introduction

This dissertation is mainly dealing with three different topics, which will turn out to be related to each other, namely polynomial matrices over finite fields, discrete-time linear systems and coding theory with focus on convolutional codes. This is reflected in the structure of this thesis, which has four chapters, where the first one provides some preliminaries and each of the remaining three chapters is devoted to one of the three topics mentioned above.

Polynomial Matrices

The mathematical importance of polynomial matrices, especially over finite fields, could be seen by the fact that they are relevant for many quite diverse mathematical areas, e.g. number theory, linear algebra, discrete-time linear systems or coding theory. Consequently, they give rise to draw relations between different mathematical fields. This is done in this dissertation by relating number theory, systems theory and coding theory.

In numerous applications of polynomial matrices certain coprimeness properties are of particular interest. The probably simplest example is a fraction of two scalar polynomials, where in most cases, it is of advantage to assume that the fraction is canceled, i.e. that numerator and denominator are coprime. Furthermore, coprimeness conditions serve as assumptions for many mathematical theorems, especially in the field of number theory, e.g. for the Chinese Remainder Theorem, to name just one of them.

In this thesis, we are mainly concerned with counting polynomial matrices having special coprimeness properties, i.e. with calculating the probability that a random polynomial matrix of a prescribed structure fulfills the corresponding coprimeness condition. The simplest case, namely the case of coprime scalar polynomials, has been investigated for a long time. It started with the observation that over the binary field, there are as many pairs of coprime polynomials as pairs of not coprime polynomials, i.e. the probability of coprimeness is equal to $1/2$. According to [15], this was firstly proven by Knuth in [26]. Generalizing this result, Bennett and Benjamin [4] computed the probability that finitely many polynomials over an arbitrary finite field are coprime by using Euclid's algorithm. An alternative and shorter proof for this result could be found in [15], where the authors additionally showed that the number of coprime polynomials is related to the number of Hankel matrices.

A polynomial matrix $A \in \mathbb{F}[z]^{n \times k}$ with $n \geq k$ is said to be right prime if $A(z_0)$ has full column rank for all z_0 from the algebraic closure $\overline{\mathbb{F}}$. Consequently, coprimeness of scalar polynomials is equivalent to the right primeness of the column vector having these polynomials as entries. Hence, right primeness of polynomial matrices could be viewed as generalization of coprimeness of polynomials. In this thesis, we will calculate the probability that a polynomial matrix with a special degree structure is right prime, whereby we will use the uniform probability distribution. Therefore, we set $t := |\mathbb{F}|^{-1}$ and investigate the asymptotic behaviour of the probability if t tends to zero, i.e. the size of the considered field becomes large. To describe the quality of our asymptotic formula, we will use the Landau symbol O: the negligible term is $O(t^e)$ for some $e \in \mathbb{N}$ if and only if it tends to zero at least as fast as t^e for $t \to 0$. If one allows the polynomial entries of the matrix to have arbitrary degrees, one could use the so-called natural density as probability measure. It was defined by Guo and Yang in [17], where the authors computed the natural density that a polynomial matrix is right prime. The proposed degree structure we use is important for the application of the resulting formula for the probability to linear systems and convolutional codes. Nevertheless, it turns out to be quite illustrative to compare the achieved formula with those for the natural density from [17].

Another possibility to generalize the notion of coprime polynomials is to consider pairwise coprime polynomials. Several authors studied pairwise coprimeness of integers, resulting in very complicated formulas; see e.g. [39], [30]. To get a formula for the probability of pairwise coprimeness for finitely many scalar polynomials, we will transfer the strategy from [30] from integers to polynomials. We will obtain a quite elaborate formula, which we will simplify by again using Landau O notation to describe the asymptotic behaviour for large field sizes. We will compare this formula with the natural density of pairwise coprime polynomials, which was computed in [18]. Since [18] was not published at the time when the research for the corresponding part of the dissertation was carried out, and not known to us until the research for the whole dissertation was completed, we also give an own proof for the natural density of pairwise coprimeness.

In fact, pairwise coprimeness is equivalent to the left primeness of a polynomial matrix that is constructed out of the involved scalar polynomials in a certain way. If one does the same construction for square polynomial matrices, i.e. constructs a bigger matrix out of them and considers its left primeness, one is actually investigating a property called mutual left coprimeness, which turns out to be a stronger condition than pairwise coprimeness in the case that the polynomial matrices are not scalar. In this thesis, we will also asymptotically calculate the probability of this property using two different probability measures.

Discrete-time Linear Systems

The original motivation for computing the probabilities of all these different coprimeness conditions were applications in the theory of discrete-time linear systems, in particular for the investigation of networks of these systems.

A discrete-time linear system over a field $\mathbb{F}$ is given by the equations

$$x(\tau + 1) = Ax(\tau) + Bu(\tau)$$
$$y(\tau) = Cx(\tau) + Du(\tau)$$

with system matrices $A \in \mathbb{F}^{n \times n}, B \in \mathbb{F}^{n \times m}, C \in \mathbb{F}^{p \times n}, D \in \mathbb{F}^{p \times m}$, input $u \in \mathbb{F}^m$, state vector $x \in \mathbb{F}^n$, output $y \in \mathbb{F}^p$ and $\tau \in \mathbb{N}_0$.

Such systems could be used to describe various phenomena in nature and engineering. If one looks at the most simplified case, namely $B = 0$ and $A = a \in \mathbb{F} \setminus \{0\}$ scalar and initially only considers the first equation, one gets

$$x(\tau + 1) = ax(t).$$

This equation describes for example the behaviour of a bacterial culture or could be used to model population growth.

If you have a matrix $B \neq 0$, it is possible to influence these natural processes via an (time varying) input u, e.g. by injecting some chemicals into the bacteria or by affecting population growth through medical developments or political decisions.

The second equation of a linear system characterizes the output y, which is a variable that is influenced by the internal parameter x as well as by the external parameter u but itself has no effect on the other parameters. There are plenty of examples for variables, which are influenced by the population and by political decisions, e.g. the amount of expended resources, to name just one of them. Surely, in long term development an increasing consumption of resources might have an impact on population growth but one has to bear in mind that no model could completely cover all aspects of reality.

In control theory, one aims to design an input u in such way that the parameter x attains some desired state. This is only possible if the matrices A and B fulfill a special property called reachability. There are two famous and frequently used tests to check reachability of linear system, developed in the middle of the 20th century, namely the so-called Kalman test and the so called Hautus test. Since the focus of this dissertation is the calculation of probabilities for certain properties over finite fields, a very important result for this work is a formula for the probability of reachability for a linear system over a finite field, provided by Helmke and others in [21].

A second important quality of a linear system is its observability. One defines a system to be observable if the knowledge of the input and output sequences is sufficient to determine the sequence of states. In the above example that would mean that one could infer the population size from the amount of resources and from political decisions. It turns out that reachability and observability are dual properties of a linear system. That means there also exist a Kalman test and a Hautus test for observability and that one could easily conclude a formula for the probability of observability from the formula for reachability.

Many complex phenomena in nature or engineering could be modelled by a network of interconnected linear systems. The main contribution of this thesis is to calculate the probability of reachability for an interconnection of linear systems. The most important tool for this is a theorem of Fuhrmann and Helmke [14], which provides criteria for reachability and observability of networks of linear systems. For the most frequently used coupling structures, namely parallel and series connection of two systems, this has already been done before by Fuhrmann ([13]), and Callier and Nahum ([5]), respectively. All these criteria could be expressed as primeness conditions on specially structured polynomial matrices. For example, to compute the probability of reachability for a parallel connection of N linear systems, one basically has to calculate the probability of mutual coprimeness for N polynomial matrices.

Coding Theory

There exists a strong relationship between discrete-time linear systems and convolutional codes because each convolutional code could be constructed out of a linear system; see [33], [35]. The basic idea of coding is to add redundancy to a message you want to send in order to make it possible to reconstruct the original message even if part of it was received incorrectly or did not arrive at all. An easy example for this is the use of check digits in various areas of life such as bank cards or ISBN numbers.

The first approach was to use block codes. In this case, encoding a message, represented by a vector of elements from a (finite) field, means multiplying it with a full column rank matrix over this field. This means the codewords are images of that matrix, which is called generator matrix of the code. One uses the Hamming distance, which is defined as the number of different components of two codewords, to measure the error correcting capability of a code. In the sequel of this thesis we will estimate the probability that a block code has optimal distance properties for error correction with respect to its parameters. This type of code is called maximum distance separable (MDS).

In 1955, convolutional codes were introduced by Elias [10]. In a certain sense, he did this generalizing the notion of block codes by using full column rank polynomial matrices as generator matrices for the code. Since then, a lot of research has been done trying to construct convolutional codes with advantageous distance properties; see e.g. [34], [25], [11]. One section of this thesis is devoted to maximum distance profile (MDP) convolutional codes, where the probability of this property is estimated.

As mentioned before, it is possible to define a convolutional code in terms of a linear system. According to [33] and [35], the obtained code is represented by the corresponding system in a minimal way if and only if the system is reachable. Moreover, if this is true, the corresponding code is non-catastrophic if and only if the system is observable. Non-catastrophicity of a code means that finitely many transmission errors could only cause finitely many decoding errors, what clearly is a very important quality. Furthermore, non-catastrophicity could be characterized by a primeness condition on a polynomial matrix. Consequently, it is possible to obtain the probability of non-catastrophicity for a convolutional code from the probability of reachability and observability of a linear system as well as by calculating the probability of the corresponding primeness condition.

Concatenating convolutional codes, i.e. considering networks of such codes, is a frequently used tool to design codes with certain properties. Again, the standard interconnections, i.e. parallel and series connection, as well as several variations and combinations of them, occur most often; see [2], [3], [7], [8], [9], [12]. Using the relationship between linear systems and convolutional codes, it is possible to transfer the reachability and observability criteria for networks of linear systems from [14] to non-catastrophicity criteria for interconnected convolutional codes. Moreover, the obtained probability results for interconnected linear systems carry over to concatenated convolutional codes.

A final type of coding that plays a role for this thesis is random linear network coding. Network coding is a quite young area of research, which was introduced by Ahlswede and others [1] in 2000 for use in wireless networks. The basic idea of linear network coding is to process information from a source via a network to several sink nodes, where each node of the network forwards a linear combination of the received symbols. In random linear network coding, one chooses these coefficients for the linear combinations randomly. The aim is that all sink nodes should finally get all source information. We will calculate the probability that this is fulfilled, i.e. the probability that one gets a solution for the network coding problem. In the final part of this dissertation, we will also look at convolutional network coding, where one has a delay between receiving and forwarding at the network nodes and consider the probability of solvability of the network coding problem in this case.

Chapter 1

Linear Systems and Polynomial Matrices

In this thesis, we want to calculate - amongst others - the probabilities for some essential properties of linear systems, such as reachability and observability; the corresponding results can be found in Chapter 3. This first chapter provides some preliminaries, which we will need for these considerations.

In Section 1.1, we start with some basic definitions and properties about discrete-time linear systems. Since polynomial matrices over finite fields play an important role when investigating such systems (see e.g. [31],[13]), we continue defining and characterizing different notions of coprimeness for polynomial matrices. Finally, this section is finished by introducing polynomial matrix fraction descriptions for the transfer function of a linear system.

In Section 1.2, we define networks of linear systems and afterwards, present and generalize criteria of Fuhrmann and Helmke [14] to characterize their reachability and observability. We conclude showing some standard examples for such networks.

1.1 Definitions and Basics

Let $\mathbb{F}$ be an arbitrary field and $A \in \mathbb{F}^{n \times n}, B \in \mathbb{F}^{n \times m}, C \in \mathbb{F}^{p \times n}, D \in \mathbb{F}^{p \times m}$. We consider discrete-time linear control systems of the form

$$x(\tau + 1) = Ax(\tau) + Bu(\tau)$$
$$y(\tau) = Cx(\tau) + Du(\tau) \tag{1.1}$$

with input $u \in \mathbb{F}^m$, state vector $x \in \mathbb{F}^n$, output $y \in \mathbb{F}^p$ and $\tau \in \mathbb{N}_0$.
In the following, we will frequently identify this system with the matrix-quadruple (A, B, C, D).

Definition 1.
A linear system (1.1) is called

(a) **reachable** *if for each $\xi \in \mathbb{F}^n$ there exist $\tau_* \in \mathbb{N}_0$ and a sequence of inputs $u(0), \ldots, u(\tau_*) \in \mathbb{F}^m$ such that the sequence of states $0 = x(0), x(1), \ldots, x(\tau_*+1)$ generated by (1.1) satisfies $x(\tau_* + 1) = \xi$.*

(b) **observable** *if $Cx(\tau)+Du(\tau) = C\tilde{x}(\tau)+Du(\tau)$ for all $\tau \in \mathbb{N}_0$ implies $x(\tau) = \tilde{x}(\tau)$ for all $\tau \in \mathbb{N}_0$. This means that the knowledge of the input and output sequences is sufficient to determine the sequence of states.*

(c) **minimal** *if it is reachable and observable.*

The following two theorems provide well-known characterizations both of reachability as well as of observability, the so-called Kalman test and the so-called Hautus test.

Theorem 1. *(Kalman test)*
A linear system (1.1) is reachable if and only if the reachability matrix
$\mathcal{R}(A, B) := (B, AB, \ldots, A^{n-1}B) \in \mathbb{F}^{n \times nm}$ *satisfies* $\mathrm{rk}(\mathcal{R}(A, B)) = n$ *and observable*

if and only if the observability matrix $\mathcal{O}(A, C) = \begin{pmatrix} C \\ \vdots \\ CA^{n-1} \end{pmatrix} \in \mathbb{F}^{pn \times n}$ *satiesfies*

$\mathrm{rk}(\mathcal{O}(A, B)) = n.$

Theorem 2. *(Hautus test)*
A linear system (1.1) is reachable if and only if $\mathrm{rk} \begin{bmatrix} zI - A & B \end{bmatrix} = n$ *for all* $z \in \overline{\mathbb{F}}$. *It is observable if and only if* $\mathrm{rk} \begin{bmatrix} zI - A \\ C \end{bmatrix} = n$ *for all* $z \in \overline{\mathbb{F}}$.

Since reachability of (1.1) only depends on the matrices A and B, one calls the pair $(A, B) \in \mathbb{F}^{n \times n} \times \mathbb{F}^{n \times m}$ reachable if and only if (1.1) is reachable. Equivalently, one calls $(A, C) \in \mathbb{F}^{n \times n} \times \mathbb{F}^{p \times n}$ observable if and only if (1.1) is observable and $(A, B, C) \in \mathbb{F}^{n \times n} \times \mathbb{F}^{n \times m} \times \mathbb{F}^{p \times n}$ minimal if and only if (1.1) is minimal.

Remark 1.
The Kalman and the Hautus test show that (A, B, C) *is observable if and only if* $(A^\top, C^\top, B^\top)$ *is reachable.*

It has been proven to be a very useful tool for the investigation of linear systems to assign a rational matrix, the so-called transfer function, to each system, which is given by the following definition:

Definition 2.
One calls the rational function $T(z) := C(zI - A)^{-1}B + D \in \mathbb{F}(z)^{p \times m}$ *the **transfer function** of system (1.1). On the other hand, the matrix-quadruple* $(A, B, C, D) \in \mathbb{F}^{n \times n} \times \mathbb{F}^{n \times m} \times \mathbb{F}^{p \times n} \times \mathbb{F}^{p \times m}$ *is called an **n-dimensional realization** of* $T(z)$. *It is called a **minimal realization** if for any other realization of* $T(z)$ *with dimension* $\tilde{n}$, *it holds* $\tilde{n} \geq n$. *In this case, one calls* $\delta(T) := n$ *the **McMillan degree** of* T.

The following theorem shows that this notion of minimality is actually equivalent to the notion of minimality from Definiton 1.

Theorem 3.
The matrix-quadruple $(A, B, C, D) \in \mathbb{F}^{n \times n} \times \mathbb{F}^{n \times m} \times \mathbb{F}^{p \times n} \times \mathbb{F}^{p \times m}$ *is a minimal realization of* $T(z) = C(zI - A)^{-1}B + D$ *if and only if it is minimal in the sense of Definition 1, i.e. if and only if it describes a reachable and observable linear system.*

It is counted among the fundamental problems of systems theory to find a realization for a given rational function. Hence, there arises the question for which functions this is even possible. Furthermore, if it is possible, one is interested in the number of such realizations and especially in the characterization of minimal realizations. The answers to these questions are contained in the following theorem, for which we need another definition as a start.

Definition 3.
*A rational function $T(z) \in \mathbb{F}(z)^{p \times m}$ is called **proper** if $\lim_{z \to \infty} T(z)$ exists and **strictly proper** if $\lim_{z \to \infty} T(z) = 0$. Equivalently, it is proper if and only if for the Laurent series expansion $T(z) = \sum_{i=-N}^{\infty} T_i z^{-i}$ with $N \in \mathbb{N}_0$ and $T_i \in \mathbb{F}^{p \times m}$ holds that $T_{-N} = \cdots = T_{-1} = 0$ and strictly proper if and only if $T_{-N} = \cdots = T_0 = 0$.*

Theorem 4.
Let $T(z) \in \mathbb{F}(z)^{p \times m}$ be a proper rational function. Then, there exists a minimal realization $(A, B, C, D) \in \mathbb{F}^{n \times n} \times \mathbb{F}^{n \times m} \times \mathbb{F}^{p \times n} \times \mathbb{F}^{p \times m}$ of $T(z)$. Moreover, if $(\tilde{A}, \tilde{B}, \tilde{C}, \tilde{D})$ is another minimal realization, there exists an invertible matrix $S \in Gl_n(\mathbb{F})$ with $A = S\tilde{A}S^{-1}$, $B = S\tilde{B}$, $C = \tilde{C}S^{-1}$, $D = \tilde{D}$.

Remark 2.
Since $\lim_{z \to \infty} C(zI - A)^{-1}B + D = D$, the transfer function of a linear system is always proper and strictly proper if and only if $D = 0$.

Scalar rational functions could be written as a quotient of polynomials, where the denominator is not allowed to be equal to zero. Moreover, there exists a reduced representation consisting of coprime polynomials, which is unique up to a constant factor. All these concepts could be generalized to rational matrices using polynomial matrix fraction descriptions, what we will do in the following.

Definition 4.
*A polynomial matrix $Q \in \mathbb{F}[z]^{m \times m}$ is called **nonsingular** if $\det(Q(z)) \not\equiv 0$. It is called **unimodular** if $\det(Q(z)) \neq 0$ for all $z \in \bar{\mathbb{F}}$, i.e. if $\det(Q(z))$ is a nonzero constant. This is the case if and only if Q is invertible in $\mathbb{F}[z]^{m \times m}$, i.e. if there exists $R(z) \in \mathbb{F}[z]^{m \times m}$ with $Q(z)R(z) = I_m$. Therefore, one denotes the group of unimodular $m \times m$-matrices over $\mathbb{F}[z]$ by $Gl_m(\mathbb{F}[z])$.*

Definition 5.
*A polynomial matrix $H \in \mathbb{F}[z]^{p \times m}$ is called a **common left divisor** of $H_i \in \mathbb{F}[z]^{p \times m_i}$ for $i = 1, \ldots, N$ if there exist matrices $X_i \in \mathbb{F}[z]^{m \times m_i}$ with $H_i(z) = H(z)X_i(z)$ for $i = 1, \ldots, N$. It is called a **greatest common left divisor**, which is denoted by $H = \gcld(H_1, \ldots, H_N)$, if for any other common left divisor $\tilde{H} \in \mathbb{F}[z]^{p \times \tilde{m}}$ there exists $S(z) \in \mathbb{F}[z]^{\tilde{m} \times m}$ with $H(z) = \tilde{H}(z)S(z)$.*
*A polynomial matrix $E \in \mathbb{F}[z]^{p \times m}$ is called a **common left multiple** of $E_i \in \mathbb{F}[z]^{m_i \times m}$ for $i = 1, \ldots, N$ if there exist matrices $X_i \in \mathbb{F}[z]^{p \times m_i}$ with $X_i(z)E_i(z) = E(z)$ for $i = 1, \ldots, N$. It is called a **least common left multiple**, which is denoted by $E = \lclm(E_1, \ldots, E_N)$, if for any other common left multiple $\tilde{E} \in \mathbb{F}[z]^{\tilde{p} \times m}$, there exists $R(z) \in \mathbb{F}[z]^{\tilde{p} \times p}$ with $R(z)E(z) = \tilde{E}(z)$.*
*One defines a **(greatest) common right divisor**, which is denoted by $\gcrd$, and a **(least) common right multiple**, which is denoted by $\lcrm$, in an analogous manner.*

Definition 6.
Let $m = m_1 + \cdots + m_N$. *Polynomial matrices* $H_i \in \mathbb{F}[z]^{p \times m_i}$ *are called* **left coprime** *if there exists* $X \in \mathbb{F}[z]^{m \times p}$ *such that* $H = \mathrm{gcld}(H_1, \ldots, H_N)$ *satisfies* $HX = I_p$. *In particular, one polynomial matrix* $H \in \mathbb{F}[z]^{p \times m}$ *is called* **left prime** *if there exists* $X \in \mathbb{F}[z]^{m \times p}$ *with* $HX = I_p$. *In an analogous manner, one defines the property to be* **right coprime** *or* **right prime**, *respectively. Note that in the case* $p = m$, *right primeness and left primeness are equivalent to the property to be unimodular.*

Since it seems not so easy to work directly with these definitions, we will use the characterizations given by the following theorem.

Theorem 5. *[14, Theorem 2.27]*
(a) *The polynomial matrices* $H_i \in \mathbb{F}[z]^{p \times m_i}$ *are left coprime if and only if*
$$\mathrm{rk}(H_1(z), \ldots, H_N(z)) = p \text{ for all } z \in \overline{\mathbb{F}}.$$

(b) *The polynomial matrices* $H_i \in \mathbb{F}[z]^{p_i \times m}$ *are right coprime if and only if*
$$\mathrm{rk} \begin{pmatrix} H_1(z) \\ \vdots \\ H_N(z) \end{pmatrix} = m \text{ for all } z \in \overline{\mathbb{F}}.$$

Remark 3.
(a) *Left coprimeness of* $H_i \in \mathbb{F}[z]^{p \times m_i}$ *is equivalent to left primeness of the matrix* $(H_1, \ldots, H_N)$ *and right coprimeness of* $H_i \in \mathbb{F}[z]^{p_i \times m}$ *is equivalent to right primeness of* $\begin{pmatrix} H_1 \\ \vdots \\ H_N \end{pmatrix}$.

(b) *A rectangular matrix* $H \in \mathbb{F}[z]^{p \times m}$ *with* $p \le m$ *is left prime if and only if its* $p \times p$-*minors are coprime. Analogously,* $H \in \mathbb{F}[z]^{p \times m}$ *with* $p \ge m$ *is right prime if and only if its* $m \times m$-*minors are coprime; see e.g. [42].*

(c) *The Hautus test shows that a pair* (A, B) *is reachable if and only if* $zI - A$ *and* B *are left coprime and that a pair* (A, C) *is observable if and only if* $zI - A$ *and* C *are right coprime.*

For some of our purposes we will need a stronger notion than just coprimeness of polynomial matrices, namely mutual coprimeness, which is characterized by the following definition.

Definition 7.
Nonsingular polynomial matrices $D_1, \ldots, D_N \in \mathbb{F}[z]^{m \times m}$ *are called* **mutually left coprime** *if for each* $i = 1, \ldots, N$, D_i *is left coprime with* $\mathrm{lcrm}\{D_j\}_{j \ne i}$.

This criterion for mutual left coprimeness is not very easy to handle. Thus, we will employ again an equivalent characterization for our later computations.

Theorem 6. *[14, Proposition 10.3]*
Nonsingular polynomial matrices $D_1, \ldots, D_N \in \mathbb{F}[z]^{m \times m}$ are mutually left coprime if and only if

$$\mathcal{D}_N := \begin{bmatrix} D_1 & D_2 & & 0 \\ & \ddots & \ddots & \\ 0 & & D_{N-1} & D_N \end{bmatrix}$$

is left prime.

Finally, we will need the following well-known characterization of coprimeness for two scalar polynomials.

Theorem 7.
Two polynomials $p(z) = \sum_{i=0}^m p_i z^i$ and $q(z) = \sum_{i=0}^n q_i z^i$ are coprime if and only if the Sylvester resultant

$$\mathrm{Res}(p,q) := \begin{bmatrix} p_0 & & q_0 & \\ \vdots & \ddots & \vdots & \ddots \\ p_m & & p_0 & q_n & & q_0 \\ & \ddots & \vdots & & \ddots & \vdots \\ & & p_m & & & q_n \end{bmatrix} \in \mathbb{F}^{(n+m) \times (n+m)}$$

is invertible.

Now, we are ready to consider coprime matrix fraction descriptions of rational matrices.

Theorem 8. *[14, Theorem 2.29]*
Let $T \in \mathbb{F}(z)^{p \times m}$ be arbitrary. Then, it holds:

(a) *There exist right coprime polynomial matrices $P \in \mathbb{F}[z]^{p \times m}$ and $Q \in \mathbb{F}[z]^{m \times m}$ nonsingular such that $T(z) = P(z)Q(z)^{-1}$.*
If $\tilde{P} \in \mathbb{F}[z]^{p \times m}$ and $\tilde{Q} \in \mathbb{F}[z]^{m \times m}$ are right coprime with $\tilde{Q}$ nonsingular such that $\tilde{P}(z)\tilde{Q}(z)^{-1} = T(z) = P(z)Q(z)^{-1}$, then there exists a (unique) unimodular matrix $U \in Gl_m(\mathbb{F}[z])$ with $\tilde{P} = PU$ and $\tilde{Q} = QU$.

(b) *There exist left coprime polynomial matrices $\hat{P} \in \mathbb{F}[z]^{p \times m}$ and $\hat{Q} \in \mathbb{F}[z]^{p \times p}$ nonsingular such that $T(z) = \hat{Q}(z)^{-1}\hat{P}(z)$.*
If $\bar{P} \in \mathbb{F}[z]^{p \times m}$ and $\bar{Q} \in \mathbb{F}[z]^{p \times p}$ are left coprime with $\bar{Q}$ nonsingular such that $\bar{Q}(z)^{-1}\bar{P}(z) = T(z) = \hat{Q}(z)^{-1}\hat{P}(z)$, then there exists a (unique) unimodular matrix $U \in Gl_m(\mathbb{F}[z])$ with $\bar{P} = \hat{P}U$ and $\bar{Q} = \hat{Q}U$.

Among this set of unimodular equivalent coprime factorizations, we focus on two particular choices, where the denominator matrix has special properties. To this end, we first need the following definitions.

Definition 8.
The *j-th column degree* of a polynomial matrix $H(z) \in \mathbb{F}[z]^{p \times m}$ is defined as $\nu_j := \deg_j H := \max_{1 \leq i \leq p} \deg(h_{ij})$. *Furthermore, let $[h_{ij}]$ denote the coefficient of z^{ν_j} in h_{ij}. Then, the **highest column degree coefficient matrix** $[H]_{hc} \in \mathbb{F}^{p \times m}$ is defined as the matrix consisting of the entries $[h_{ij}]$. For $p = m$, one calls H **column proper** if $[H]_{hc} \in Gl_m(\mathbb{F})$.*

Definition 9. [14, Corollary 2.42], [22, Proposition 5.1]
Let $Q \in \mathbb{F}[z]^{m \times m}$ be nonsingular. Then, there exist

(a) *a unimodular matrix $U_1 \in Gl_m(\mathbb{F}[z])$ such that*

$$
QU_1 = Q^H := \begin{bmatrix} q_{11}^{(H)} & 0 & \cdots & 0 \\ \vdots & \ddots & \ddots & \vdots \\ \vdots & & \ddots & 0 \\ q_{m1}^{(H)} & \cdots & \cdots & q_{mm}^{(H)} \end{bmatrix}
$$

*with $q_{ii}^{(H)}$ monic and $\deg q_{ij}^{(H)} < \deg q_{ii}^{(H)} =: \kappa_{m+1-i}$ for $1 \leq j < i \leq m$. Q^H is unique and is called **Hermite canonical form**. Moreover, Q^H is called of **simple form** if $\kappa_j = 0$ for $j \geq 2$.*

(b) *a unimodular matrix $U_2 \in Gl_m(\mathbb{F}[z])$ such that*

$$
QU_2 = Q^{KH} := \begin{bmatrix} q_{11}^{(KH)} & \cdots & q_{1m}^{(KH)} \\ \vdots & & \vdots \\ q_{m1}^{(KH)} & \cdots & q_{mm}^{(KH)} \end{bmatrix}
$$

*with $q_{ii}^{(KH)}$ monic, $\deg q_{ij}^{(KH)} < \deg q_{ii}^{(KH)}$ for $j \neq i$, $\deg q_{ji}^{(KH)} < \deg q_{ii}^{(KH)}$ for $j < i$ and $\deg q_{ji}^{(KH)} \leq \deg q_{ii}^{(KH)}$ for $j > i$. Q^{KH} is unique and is called **Kronecker-Hermite canonical form**. Note that it is always column proper.*

Theorem 9.
Let $T(z) = P(z)Q(z)^{-1}$ with $P \in \mathbb{F}[z]^{p \times m}, Q \in \mathbb{F}[z]^{m \times m}, \det(Q) \not\equiv 0$ a right coprime factorization of the transfer function. Then, it holds:

(a) $\delta(T) = \deg(\det(Q))$.

(b) *For every unimodular matrix $U \in Gl_m(\mathbb{F}[z])$, the pair (PU, QU) is also a right coprime factorization of the corresponding transfer function. Consequently, one could either assume that $Q = Q^{KH}$ or that $Q = Q^H$.*

Proof.

(a) This equality is one of the statements of Theorem 4.24 of [14].

(b) Since Q is nonsingular, there exist (unique) unimodular matrices U_1 und U_2, such that QU_1 is in Kronecker-Hermite and QU_2 is in Hermite canonical form. If one takes the pair (PU_i, QU_i) instead of (P, Q) for $i = 1, 2$, one still has a right coprime factorization of the transfer function because $(PU_i)(QU_i)^{-1} = PU_iU_i^{-1}Q^{-1} = PQ^{-1}$ and multiplying with unimodular matrices does not change coprimeness.

$\square$

So far, we only focused on the structure of the denominator matrix Q. But if it is in Kronecker-Hermite from, i.e. especially column proper, one also has some knowledge about the numerator matrix P .

Lemma 1. *[14, Proposition 2.30]*
Let $(A, B, C, D) \in \mathbb{F}^{n \times n} \times \mathbb{F}^{n \times m} \times \mathbb{F}^{p \times n} \times \mathbb{F}^{p \times m}$ and $C(zI - A)^{-1}B + D = P(z)Q(z)^{-1}$ with $P \in \mathbb{F}[z]^{p \times m}, Q \in \mathbb{F}[z]^{m \times m}, \det(Q) \neq 0$ and Q column proper. Then, one has for $j = 1, ..., m$:

$$\deg_j P(z) \leq \deg_j Q(z) \quad and \quad \deg_j P(z) < \deg_j Q(z) \text{ if } D = 0.$$

1.2 Networks of Linear Systems

With the help of polynomial matrix fraction descriptions for linear systems, which we introduced in the preceding subsection, we are now able to analyze interconnections of linear systems. Doing this, we consider a network of N linear systems defined by the following equations

$$x_i(\tau + 1) = A_i x_i(\tau) + B_i v_i(\tau) \tag{1.2}$$
$$w_i(\tau) = C_i x_i(\tau) + D_i v_i(\tau) \tag{1.3}$$
$$v_i(\tau) = \sum_{j=1}^{N} K_{ij} w_j(\tau) + L_i u(\tau) \tag{1.4}$$
$$y(\tau) = \sum_{i=1}^{N} M_i w_i(\tau) + J u(\tau) \tag{1.5}$$

with $A_i \in \mathbb{F}^{n_i \times n_i}, B_i \in \mathbb{F}^{n_i \times m_i}, C_i \in \mathbb{F}^{p_i \times n_i}, D_i \in \mathbb{F}^{p_i \times m_i}, K_{ij} \in \mathbb{F}^{m_i \times p_j}, L_i \in \mathbb{F}^{m_i \times m}, M_i \in \mathbb{F}^{p \times p_i}$ and $J \in \mathbb{F}^{p \times m}$.
Equations (1.2) and (1.3) describe the dynamics at the single systems, which are called **node systems**. Equation (1.4) gives information about the interconnection structure between the node systems and equation (1.5) provides a formula for the output of the whole network.
In the following, assume that (A_i, B_i, C_i) are minimal for $i = 1, \ldots, N$.

Define $K := (K_{ij})_{ij}$, $L := \begin{bmatrix} L_1 \\ \vdots \\ L_N \end{bmatrix}$ and $M := [M_1, \ldots, M_N]$ as well as

$$A := \begin{bmatrix} A_1 & & 0 \\ & \ddots & \\ 0 & & A_N \end{bmatrix}, B := \begin{bmatrix} B_1 & & 0 \\ & \ddots & \\ 0 & & B_N \end{bmatrix}, C := \begin{bmatrix} C_1 & & 0 \\ & \ddots & \\ 0 & & C_N \end{bmatrix} \text{ and}$$

$$D := \begin{bmatrix} D_1 & & 0 \\ & \ddots & \\ 0 & & D_N \end{bmatrix}.$$

Thus, the global state space representation of the decoupled node systems is

$$x(\tau + 1) = Ax(\tau) + Bv(\tau)$$
$$w(\tau) = Cx(\tau) + Dv(\tau)$$

and the interconnection is

$$v(\tau) = Kw(\tau) + Lu(\tau)$$
$$y(\tau) = Mw(\tau) + Ju(\tau).$$

To get the interconnected system into the form

$$x(\tau + 1) = \mathcal{A}x(\tau) + \mathcal{B}u(\tau)$$
$$y(\tau) = \mathcal{C}x(\tau) + \mathcal{D}u(\tau),$$

with $(\mathcal{A}, \mathcal{B}, \mathcal{C}, \mathcal{D}) \in \mathbb{F}^{\bar{n} \times \bar{n}} \times \mathbb{F}^{\bar{n} \times m} \times \mathbb{F}^{p \times \bar{n}} \times \mathbb{F}^{p \times m}$ where $\bar{n} = \sum_{i=1}^{N} n_i$, one has to write $w(\tau)$ as a linear combination of $x(\tau)$ and $u(\tau)$. Inserting the expression for $v(\tau)$ into the equation for $w(\tau)$, leads to $(I - DK)w(t) = Cx(\tau) + DLu(\tau)$. Therefore, in the following, one only considers interconnection structures with $\det(I - DK) \neq 0$. In this case, one obtains $w(\tau) = (I - DK)^{-1}Cx(\tau) + (I - DK)^{-1}DLu(\tau)$ and

$$\mathcal{A} = A + BK(I - DK)^{-1}C$$
$$\mathcal{B} = BK(I - DK)^{-1}DL + BL = B(I - KD)^{-1}L$$
$$\mathcal{C} = M(I - DK)^{-1}C$$
$$\mathcal{D} = M(I - DK)^{-1}DL + J.$$

For the transformation of $\mathcal{B}$, the identity $(K(I - DK)^{-1}D + I)(I - KD) = K(I - DK)^{-1}(I - DK)D + I - KD = I$ was used.

Finally, for $i = 1, \ldots, N$, consider right and left coprime factorizations

$$C_i(zI - A_i)^{-1}B_i + D_i = P_i(z)Q_i^{-1}(z) = \hat{Q}_i(z)^{-1}\hat{P}_i(z)$$

and define $P(z) := \begin{bmatrix} P_1(z) & & 0 \\ & \ddots & \\ 0 & & P_N(z) \end{bmatrix}$, $Q(z) := \begin{bmatrix} Q_1(z) & & 0 \\ & \ddots & \\ 0 & & Q_N(z) \end{bmatrix}$. $\hat{Q}$

and $\hat{P}$ are defined analogously.

Then, $\mathcal{C}(zI - \mathcal{A})^{-1}\mathcal{B} + \mathcal{D} = MP(z)(Q(z) - KP(z))^{-1}L + J =$
$= M(\hat{Q}(z) - \hat{P}(z)K)^{-1}\hat{P}(z)L + J$; see [14, p. 465].

The following theorem generalizes part of Theorem 9.8 of [14] to the case that the interconnected systems are just proper and not necessarily strictly proper as required there.

Theorem 10.
With the same notations as above and with the assumptions that (A_i, B_i, C_i, D_i) are minimal for $i = 1, \dots, N$ and $\det(I - DK) \neq 0$, one has:

1. *The following statements are equivalent:*

   (a) *$(\mathcal{A}, \mathcal{B})$ is reachable.*

   (b) *$Q(z) - KP(z), L$ are left coprime.*

   (c) *$\hat{Q}(z) - \hat{P}(z)K, \hat{P}(z)L$ are left coprime.*

2. *The following statements are equivalent:*

   (a) *$(\mathcal{A}, \mathcal{C})$ is observable.*

   (b) *$Q(z) - KP(z), MP(z)$ are right coprime.*

   (c) *$\hat{Q}(z) - \hat{P}(z)K, M$ are right coprime.*

Proof.
We follow the lines of the proof of Theorem 9.8 in [14] and start with the equivalence of (a) and (b) in both parts of the theorem to be proven. According to Definition 4.30, Theorem 4.26 and Theorem 4.32 of [14], one has to show that there exist appropriately sized polynomial matrices $E(z)$, $F(z)$, $X(z)$, $Y(z)$ with $E(z), zI - \mathcal{A}$ left coprime and $Q(z) - KP(z), F(z)$ right coprime such that

$$
\begin{bmatrix} E(z) & 0 \\ -X(z) & I \end{bmatrix} \cdot \begin{bmatrix} Q(z) - KP(z) & -L \\ MP(z) & J \end{bmatrix} = \begin{bmatrix} zI - \mathcal{A} & -\mathcal{B} \\ \mathcal{C} & \mathcal{D} \end{bmatrix} \cdot \begin{bmatrix} F(z) & Y(z) \\ 0 & I \end{bmatrix}.
$$

The above equation is equivalent to the following four equations:

$$E(z)(Q(z) - KP(z)) = (zI - A - BK(I - DK)^{-1}C)F(z) \tag{1.6}$$

$$-E(z)L = (zI - A - BK(I - DK)^{-1}C)Y(z) - B(I - KD)^{-1}L \tag{1.7}$$

$$-X(z)(Q(z) - KP(z)) + MP(z) = M(I - DK)^{-1}CF(z) \tag{1.8}$$

$$X(z)L + J = M(I - DK)^{-1}CY(z) + M(I - DK)^{-1}DL + J \tag{1.9}$$

Define $Y(z) := 0$, $E(z) := B(I - KD)^{-1} = BK(I - DK)^{-1}D + B$,
$X(z) := M(I - DK)^{-1}D$ and $F(z) := (zI - A)^{-1}BQ(z)$.
One immediately sees that these matrices fulfill (1.7) and (1.9). Equation (1.8) gets the form

$$-M(I - DK)^{-1}D(Q(z) - KP(z)) + MP(z) = M(I - DK)^{-1}C(zI - A)^{-1}BQ(z),$$

which is fulfilled if

$$- D(Q(z) - KP(z)) + (I - DK)P(z) = C(zI - A)^{-1}BQ(z) =$$
$$= (P(z)Q(z)^{-1} - D)Q(z) = P(z) - DQ(z),$$

which clearly is true. Equation (1.6) is equivalent to

$$B(I - KD)^{-1}(Q(z) - KP(z)) =$$
$$= (zI - A - BK(I - DK)^{-1}C)(zI - A)^{-1}BQ(z)$$
$$\Leftrightarrow BK(I - DK)^{-1}DQ(z) - (BK(I - DK)^{-1}D + B)KP(z)) =$$
$$= -BK(I - DK)^{-1}C(zI - A)^{-1}BQ(z).$$

This is true if

$$DQ(z) - DKP(z) - (I - DK)P(z) = -C(zI - A)^{-1}BQ(z),$$

which was already shown considering equation (1.8).

It remains to show that $F(z)$ is a polynomial, which is left coprime with $Q(z) - KP(z)$, as well as that $zI - A, E(z)$ are right coprime.

Therefore, consider the coprime factorization $\tilde{P}(z)\tilde{Q}(z)^{-1} = (zI - A)^{-1}B$. With the definitions $\tilde{Y}(z) := 0$, $\tilde{E}(z) := B$, $\tilde{X}(z) := D$, $\tilde{F}(z) := \tilde{P}(z)$ and $H(z) := C\tilde{P}(z) + D\tilde{Q}(z)$, one has

$$\begin{bmatrix} \tilde{E}(z) & 0 \\ -\tilde{X}(z) & I \end{bmatrix} \cdot \begin{bmatrix} \tilde{Q}(z) & -I \\ H(z) & 0 \end{bmatrix} = \begin{bmatrix} zI - A & -B \\ C & D \end{bmatrix} \cdot \begin{bmatrix} \tilde{F}(z) & \tilde{Y}(z) \\ 0 & I \end{bmatrix},$$

where $\tilde{E} = B$ and $zI - A$ are left coprime because the single systems are reachable and $\tilde{Q}(z)$ and $\tilde{F}(z)$ are right coprime per definition. Furthermore, as the node systems are minimal, $zI - A$ and B are left coprime and $zI - A$ and C are right coprime. It follows from Definition 4.30, Theorem 4.26 and Theorem 4.32 of [14] that $H(z)\tilde{Q}(z)^{-1}$ is a right coprime factorization of $C(zI - A)^{-1}B + D$. Therefore, there exists a unimodular matrix $U(z)$ with $P(z) = H(z)U(z)$ and $Q(z) = \tilde{Q}(z)U(z)$ and thus, $F(z) = (zI - A)^{-1}BQ(z) = \tilde{P}(z)U(z)$ is a polynomial. Moreover, since $Q(z)$ and $P(z)$ are right coprime per definition and $\det(I - DK) \neq 0$ per assumption,

$$\begin{bmatrix} I & 0 \\ 0 & C \end{bmatrix} \cdot \begin{bmatrix} Q(z) - KP(z) \\ (zI - A)^{-1}BQ(z) \end{bmatrix} = \begin{bmatrix} Q(z) - KP(z) \\ (P(z)Q(z)^{-1} - D)Q(z) \end{bmatrix} =$$
$$= \begin{bmatrix} I & -K \\ -D & I \end{bmatrix} \cdot \begin{bmatrix} Q(z) \\ P(z) \end{bmatrix}.$$

has full column rank, i.e. the corresponding map is injective, for all $z \in \overline{\mathbb{F}}$. Therefore, the map defined by $\begin{bmatrix} Q(z) - KP(z) \\ (zI - A)^{-1}BQ(z) \end{bmatrix}$ has to be injective, i.e. of full column rank, for all $z \in \overline{\mathbb{F}}$, too. Consequently, $Q(z) - KP(z), F(z)$ are right coprime.

That the other coprimeness condition is fulfilled as well follows from

$\mathrm{rk}[B(I-KD)^{-1} \quad zI-A-BK(I-DK)^{-1}C] =$

$= \mathrm{rk}[B(I-KD)^{-1}$

$\quad zI-A-BK(I-DK)^{-1}C + B(I-KD)^{-1}(I-KD)K(I-DK)^{-1}C] =$

$= \mathrm{rk}[B(I-KD)^{-1} \quad zI-A] =$

$= \mathrm{rk}\left([B(I-KD)^{-1} \quad zI-A] \cdot \begin{bmatrix} I-KD & 0 \\ 0 & I \end{bmatrix} \right) =$

$= \mathrm{rk}[B \quad zI-A]$

and the fact that the single systems are reachable per assumption.

Finally, the equivalences between (b) and (c) in both parts of the theorem hold because $\hat{Q}(z)^{-1}\hat{P}(z) = P(z)Q(z)^{-1}$ implies

$$\begin{bmatrix} \hat{P}(z) & 0 \\ 0 & I \end{bmatrix} \cdot \begin{bmatrix} Q(z) - KP(z) & -L \\ MP(z) & J \end{bmatrix} =$$
$$= \begin{bmatrix} \hat{Q}(z) - \hat{P}(z)K & -\hat{P}(z)L \\ M & J \end{bmatrix} \cdot \begin{bmatrix} P(z) & 0 \\ 0 & I \end{bmatrix},$$

where $\hat{P}(z)$ and $\hat{Q}(z) - \hat{P}(z)K$ are left coprime since $\hat{P}(z)$ and $\hat{Q}(z)$ are left coprime and $P(z)$ and $Q(z) - KP(z)$ are right coprime since $P(z)$ and $Q(z)$ are right coprime. $\square$

Remark 4.
In the preceding theorem, it was assumed that the node systems are minimal. But this assumption is necessary anyway since reachability/observability of the interconnected system implies that the node systems are reachable/observable.

Proof.
$(\mathcal{A}, \mathcal{B})$ is reachable if and only if $[zI-A-BK(I-DK)^{-1}C \quad B(I-KD)^{-1}L] =$
$[zI-A-BK(I-DK)^{-1}C \quad B(I-KD)^{-1}] \cdot \begin{bmatrix} I & 0 \\ 0 & L \end{bmatrix}$ is surjective for all $z \in \bar{\mathbb{F}}$.
This implies that $[zI-A-BK(I-DK)^{-1}C \quad B(I-KD)^{-1}]$ is surjective, i.e. of full row rank, for all $z \in \bar{\mathbb{F}}$. As done at the end of the previous proof, one could show that the rank of this matrix equals the rank of $[zI-A \quad B]$. Hence, the diagonal structure of A and B effects that the node systems are reachable.

Analogously, $(\mathcal{A}, \mathcal{C})$ is observable if and only if

$$\begin{bmatrix} zI-A-BK(I-DK)^{-1}C \\ M(I-DK)^{-1}C \end{bmatrix} =$$
$$= \begin{bmatrix} I & 0 \\ 0 & M(I-DK)^{-1} \end{bmatrix} \cdot \begin{bmatrix} zI-A-BK(I-DK)^{-1}C \\ C \end{bmatrix}$$

is injective for all $z \in \bar{\mathbb{F}}$. This implies that $\begin{bmatrix} zI-A-BK(I-DK)^{-1}C \\ C \end{bmatrix}$ is injective, i.e. of full column rank, for all $z \in \bar{\mathbb{F}}$. But the rank of this matrix is equal to the rank of $\begin{bmatrix} zI - A \\ C \end{bmatrix}$. Again, the diagonal structure of A and C effects the observability of the node systems. □

Corollary 1.
An interconnected system (A, B, C, D) is minimal if and only if the node systems are minimal, $Q(z) - KP(z), L$ are left coprime and $Q(z) - KP(z), MP(z)$ are right coprime or equivalently, if the node systems are minimal, $\hat{Q}(z) - \hat{P}(z)K, \hat{P}(z)L$ are left coprime and $\hat{Q}(z) - \hat{P}(z)K, M$ are right coprime.

In the following, the achieved criteria should be applied to some standard interconnection structures.

Example 1.

(a) *Parallel connection*

Here, one has $K = 0$, $L^\top = [I_m \ldots I_m]$ and $M = [I_p \ldots I_p]$. Hence, $I - DK$ is invertible for all values of D. Applying the preceding theorem, one gets that this

interconnection is reachable if and only if $\begin{bmatrix} Q_1(z) & & I_m \\ & \ddots & \vdots \\ & & Q_N(z) & I_m \end{bmatrix}$ is left

prime, which is the

case if and only if $\begin{bmatrix} Q_1(z) & -Q_2(z) & \\ & \ddots & \ddots \\ & & Q_{N-1}(z) & -Q_N(z) \end{bmatrix}$ is left prime, i.e. if

and only if $Q_1, \ldots, Q_N$ are mutually left coprime. This could be seen by adding the second block of rows to the first, afterwards the third block of rows to the second,

and so on. Equivalent is the condition that $\begin{bmatrix} \hat{Q}_1(z) & & \hat{P}_1(z) \\ & \ddots & \vdots \\ & & \hat{Q}_N(z) & \hat{P}_N(z) \end{bmatrix}$ is

left prime. The

interconnection is observable if and only if $\begin{bmatrix} Q_1(z) & & \\ & \ddots & \\ & & Q_N(z) \\ P_1(z) & \ldots & P_N(z) \end{bmatrix}$ is right prime

or equivalently, $\begin{bmatrix} \hat{Q}_1(z) & & \\ & \ddots & \\ & & \hat{Q}_N(z) \\ I_p & \ldots & I_p \end{bmatrix}$ is right prime, which is equivalent to

mutual left coprimeness of $\hat{Q}_1^\top, \ldots, \hat{Q}_N^\top$.

(b) Series connection

In this case, it holds $K = \begin{bmatrix} 0 & \cdots & & \cdots & 0 \\ I_{p_1} & \ddots & & & \vdots \\ & \ddots & \ddots & & \vdots \\ 0 & & I_{p_{N-1}} & 0 \end{bmatrix}$, $L = \begin{bmatrix} I_m \\ 0 \\ \vdots \\ 0 \end{bmatrix}$ *and*

$M = [0 \cdots 0 \; I_{p_N}]$.

One has $\det(I - DK) = 1 \neq 0$ *and therefore, Theorem 10 could be applied for all possible values of* D. *Doing this, one gets that the interconnection is reachable if and only if*

$$\begin{bmatrix} Q_1(z) & & & & I_m \\ -P_1(z) & Q_2(z) & & & 0 \\ & \ddots & \ddots & & \vdots \\ & & -P_{N-1}(z) & Q_N(z) & 0 \end{bmatrix}$$ *is left prime, which is the case if*

and

only if $\begin{bmatrix} -P_1(z) & Q_2(z) & & \\ & \ddots & \ddots & \\ & & -P_{N-1}(z) & Q_N(z) \end{bmatrix}$ *is left prime. Equivalent is the*

condition that $\begin{bmatrix} \hat{Q}_1(z) & & & & \hat{P}_1(z) \\ -\hat{P}_2(z) & \hat{Q}_2(z) & & & 0 \\ & \ddots & \ddots & & \vdots \\ & & -\hat{P}_N(z) & \hat{Q}_N(z) & 0 \end{bmatrix}$ *is left prime. The*

interconnection is observable if and only if

$$\begin{bmatrix} Q_1(z) & & & \\ -P_1(z) & Q_2(z) & & \\ & \ddots & \ddots & \\ & & -P_{N-1}(z) & Q_N(z) \\ & & & -P_N(z) \end{bmatrix}$$ *is right prime or equivalently,*

$$\begin{bmatrix} \hat{Q}_1(z) & & \\ -\hat{P}_2(z) & \ddots & \\ & \ddots & \hat{Q}_{N-1}(z) \\ & & -\hat{P}_N(z) \end{bmatrix}$$ *is right prime.*

(c) Circular/Feedback interconnection

Here, one has $K = \begin{bmatrix} 0 & \cdots & 0 & I_{p_N} \\ I_{p_1} & \ddots & & 0 \\ & \ddots & \ddots & \vdots \\ 0 & & I_{p_{N-1}} & 0 \end{bmatrix}$, $L = \begin{bmatrix} I_m \\ 0 \\ \vdots \\ 0 \end{bmatrix}$ *and* $M = [I_{p_1} \; 0 \cdots 0]$.

Thus, Theorem 10 could be applied if

$$
\det(I - DK) = \det \left(\begin{bmatrix} I & \cdots & 0 & -D_1 \\ -D_2 & \ddots & & 0 \\ & \ddots & \ddots & \vdots \\ 0 & & -D_N & I \end{bmatrix} \right) =
$$

$$
= \det \left(\begin{bmatrix} I & 0 & \cdots & 0 \\ -D_3 & \ddots & \ddots & \vdots \\ & \ddots & \ddots & 0 \\ 0 & & -D_N & I \end{bmatrix} - \begin{bmatrix} -D_2 \\ 0 \\ \vdots \\ 0 \end{bmatrix} \cdot [0 \ldots 0 - D_1] \right) =
$$

$$
= \det \left(\begin{bmatrix} I & \cdots & 0 & -D_2 \cdot D_1 \\ -D_3 & \ddots & & 0 \\ & \ddots & \ddots & \vdots \\ 0 & & -D_N & I \end{bmatrix} \right) = \cdots =
$$

$$
= \det \left(\begin{bmatrix} I & -D_{N-1} \cdots D_1 \\ -D_N & I \end{bmatrix} \right) = \det(I - D_N \cdots D_1) \neq 0.
$$

For this computation, the formula $\det \begin{bmatrix} I & A \\ B & C \end{bmatrix} = \det(C - BA)$ was used several times.

With the above assumption, the interconnection is reachable if and only if

$$
\begin{bmatrix} Q_1(z) & & & -P_N(z) & I_m \\ -P_1(z) & Q_2(z) & & & 0 \\ & \ddots & \ddots & & \vdots \\ & & -P_{N-1}(z) & Q_N(z) & 0 \end{bmatrix}
$$

is left prime, which clearly is equivalent to the reachability criterion for series connection. Moreover, the interconnection is observable if and only if

$$
\begin{bmatrix} Q_1(z) & & & -P_N(z) \\ -P_1(z) & Q_2(z) & & \\ & \ddots & \ddots & \\ & & -P_{N-1}(z) & Q_N(z) \\ & & & P_N(z) \end{bmatrix}
$$

is right prime, which is equivalent to the criterion for series connection, too. In a similiar way, one can show that the reachability and observability criteria using $\hat{Q}$ and $\hat{P}$ are equivalent for series and circular interconnection.

Chapter 2

Counting Problems

To compute the probability that a mathematical object has a special property, it is necessary to count mathematical objects. Therefore, in the following, we restrict our considerations to a finite field $\mathbb{F}$, which is endowed with the uniform probability distribution that assigns to each field element the same probability

$$t = \frac{1}{|\mathbb{F}|}$$

and denote the corresponding probability of a set A by $\Pr(A)$.

In addition to computing probabilities with the uniform distribution, which is only defined for finite sets, we will compare these results with the results one gets using another definition of probability, namly the concept of natural density as defined in [17] for infinite sets:

Definition 10.
To enumerate $\mathbb{F}[z]$, assign the number $k = \sum_{i=0}^{\infty} a_i (\frac{1}{t})^i$ to the polynomial $f_k(z) = \sum_{i=0}^{\infty} a_i z^i$, where $a_i = 0$ for all but finitely many coefficients and for the computation of k, the a_i are considered as elements of $\mathbb{Z}$. In particular, $f_0 \equiv 0$. Moreover, let $\mathcal{M}_n$ be the set of tuples $(D_1, \ldots, D_N) \in (\mathbb{F}[z]^{l \times m})^N$ for which the entries of D_i are elements of the set $\{f_0, \ldots, f_n\}$ for $i = 1, \ldots, N$.
The natural density of a set $E \subset (\mathbb{F}[z]^{l \times m})^N$ in $(\mathbb{F}[z]^{l \times m})^N$ is defined as $\lim_{n \to \infty} \frac{|E \cap \mathcal{M}_n|}{|\mathcal{M}_n|}$ if this limit exists.

In this chapter, we count polynomial matrices (over finite fields) fulfilling special coprimeness conditions since there number enables us to calculate the probabilities of reachability (and observability) for networks of linear systems by using Theorem 10.

In the first section of this chapter, we will present several already known results concerning general counting strategies and cardinality formulas as well as a few own extentions of these formulas and some newly developed counting methods.

In Section 2.2, we calculate the probability that a rectangular matrix of a special form is right prime because this value plays an important role for calculating the probability of minimality for a single system in Section 3.1.

The remaining and main part of this chapter is devoted to the calculation of the probability of mutual left coprimeness as this probability is crucial for the probability that a parallel connection is reachable; see Example 1 (a). In Section 2.3, this is firstly done for the case of scalar polynomials (where mutual and pairwise coprimeness coincide), which corresponds to a parallel connection of single-input systems. Finally, in Section 2.4, we do the same for the general case of square matrices with arbitrary sizes.

The results of Sections 2.2, 2.3 and 2.4 are compared with the formulas one gets using the natural density instead of the uniform probability distribution at the end of the corresponding paragraphs.

2.1 General Counting Strategies

In this first subsection, we present some general counting results, which we will need for our purposes. Some of them are well-known and therefore, stated without proof, while others are newly developed and hence, require a proof.

We start with a fundamental and frequently used principle for the determination of cardinalities, namely the so-called inclusion-exclusion principle.

Lemma 2. *(Inclusion-Exclusion Principle)*
Let $A_1, \ldots, A_n$ be finite sets and $X = \bigcup_{i=1}^{n} A_i$. For $I \subset \{1, \ldots, n\}$, define $A_I := \bigcap_{i \in I} A_i$. Then, it holds

$$|X| = \sum_{\emptyset \neq I \subset \{1, \ldots, n\}} (-1)^{|I|-1} |A_I|.$$

Another famous result is the Schwartz-Zippel lemma, which makes it possible to estimate the cardinality of varieties over finite fields.

Lemma 3. *(Schwartz-Zippel)* [36, Corollary 1]
Let $f \in \mathbb{F}[x_1, \ldots, x_r]$ be a nonzero polynomial of total degree d. Moreover, let $v_1, \ldots, v_r$ be selected at random independently and uniformly from $\mathbb{F}$. Then

$$\Pr(f(v_1, \ldots, v_r) = 0) \leq d \cdot t.$$

In Section 2.2, we will consider right prime polynomial matrices. For the case that these are constant, this property simply means being of full rank and there already exists a formula for the corresponding cardinality as the following lemma shows.

Lemma 4. *[27, S. 455]*
For $1 \leq r \leq \min(k, n)$, denote by $N(k, n, r)$ the number of matrices from $\mathbb{F}^{k \times n}$ with rank r. Then, it holds

$$N(k, n, r) = t^{-nr} \prod_{j=n-r+1}^{n} (1 - t^j) \cdot \prod_{i=0}^{r-1} \frac{t^{i-k} - 1}{t^{-(i+1)} - 1}.$$

In particular, the number of invertible $n \times n$-matrices over $\mathbb{F}$ is equal to

$$|Gl_n(\mathbb{F})| = t^{-n^2} \prod_{j=1}^{n} (1 - t^j).$$

Next, we consider another special case of polynomial matrices, namly matrices with only one row or column, i.e. vectors of polynomials. Here, being of full rank is equivalent to the fact that the polynomial entries are coprime, whose probability has also already been computed before:

Lemma 5. *[15, Theorem 4.1]*
The probability that N monic polynomials $d_1, \ldots, d_N \in \mathbb{F}[z]$ with $\deg(d_i) = n_i \in \mathbb{N}$ for $i = 1, \ldots, N$ are coprime is equal to $1 - t^{N-1}$.

Remark 5.
In [15], directly before stating the preceding theorem, the authors remarked that one gets the same probability when considering arbitrary (not necessarily monic) polynomials. This clearly is true as long as the polynomials have a fixed nonzero degree. However, at some points, we will need the probability of coprimeness in the case that the degree of several of the polynomials is only upper bounded, which allows them to be constant. Therefore, we require the following considerations.

Corollary 2.
For $N \in \mathbb{N}$ and $M \in \mathbb{N}_0$, the probability that $d_1, \ldots, d_{N+M} \in \mathbb{F}[z]$, where d_i monic with $\deg(d_i) = n_i \in \mathbb{N}$ for $i = 1, \ldots, N$ and $\deg(d_i) \leq n_i \in \mathbb{N}_0$ for $i = N+1, \ldots, N+M$, are coprime is equal to $1 - t^{N+M-1}$.

Proof.
This corollary is shown per induction with respect to M. For $M = 0$, the statement follows from Lemma 5. Assume the statement is true for M not necessarily monic polynomials. If d_{N+M+1} is the zero polynomial, which occurs with probability $t^{n_{N+M+1}+1}$, the $N + M + 1$ polynomials are coprime if and only if the first $N + M$ polynomials are coprime, which has a probability of $1 - t^{N+M-1}$ per induction. If d_{N+M+1} is a nonzero constant, which happens with probability $t^{n_{N+M+1}+1}(t^{-1} - 1)$, the polynomials are always coprime. Finally, if d_{N+M+1} is not constant, which appears with probability $\frac{t^{-(n_{N+M+1}+1)} - t^{-1}}{t^{-(n_{N+M+1}+1)}}$, it follows per induction and from Remark 5 that the proability of coprimeness is equal to $1 - t^{N+M}$. Summing up all these cases, one gets that the overall probability is equal to

$$t^{n_{N+M+1}+1}(1 - t^{N+M-1}) + t^{n_{N+M+1}+1}(t^{-1} - 1) + \frac{t^{-(n_{N+M+1}+1)} - t^{-1}}{t^{-(n_{N+M+1}+1)}}(1 - t^{N+M})$$

$$= t^{n_{N+M+1}+1}(t^{-1} - t^{N+M-1}) + (1 - t^{n_{N+M+1}})(1 - t^{N+M}) = 1 - t^{N+M}.$$

$\square$

Remark 6.
The statement of the preceding corollary is not true if $N = 0$, i.e. if there is no monic polynomial involved. To see this, consider for example the case $n_1 = n_2 = 1$ over the binary field, i.e. $t = 1/2$. Listing all possible pairs of polynomials of degree at most 1 and counting the coprime pairs, one gets that the probability of being coprime is equal to $\frac{9}{16} \neq 1 - \frac{1}{2}$.

To be able to estimate the probability for different coprimeness conditions in further parts of this work, we need to estimate the number of polynomials whose value at a given point z_0 is fixed. Since this implies considering the minimal polynomial of z_0 over $\mathbb{F}$ and the minimal polynomials of elements of $\overline{\mathbb{F}}$ are exactly the irreducible polynomials over $\mathbb{F}$, we will need the following well-known result counting irreducible polynomials of a certain degree.

Lemma 6.
The number of monic irreducible polynomials in $\mathbb{F}[z]$ of degree j is equal to

$$\varphi_j = \frac{1}{j}\sum_{d|j}\mu(d)t^{-j/d} = \frac{1}{j}t^{-j} + O(t^{-(j-1)})$$

where for $n \in \mathbb{N}$, $\mu(n) := \begin{cases}(-1)^{|\{p\in\mathbb{P} \;|\; p|n\}|}, & n \text{ square-free}\\ 0, & \text{otherwise}\end{cases}$.

Remark 7.
For $z_0 \in \overline{\mathbb{F}}$, denote by f_{z_0} its minimal polynomial over $\mathbb{F}$ and set $g_{z_0} := \deg(f_{z_0})$. Then, for $g \in \mathbb{N}$, the number of $z_0 \in \overline{\mathbb{F}}$ with $g_{z_0} = g$ is at most $\varphi_g \cdot g = O(t^g)$. In particular, for $g = 1$, it is equal to t.

Proof.
The statement follows directly from the fact that there are φ_g possible minimal polynomials for z_0 and each of them has at most g different zeros. $\qquad\square$

Lemma 7.
Let $z_0, z_1 \in \overline{\mathbb{F}}$ with $z_0 \neq z_1$ as well as $n \in \mathbb{N}$ be fixed. Then, it holds:

(a) *The number of $d \in \mathbb{F}[z]$ monic with $\deg(d) = n$ such that $d(z_0) = 0$ is equal to $t^{-n+g_{z_0}}$ if $n \geq g_{z_0}$ and zero if $n < g_{z_0}$. Moreover, the number of $d \in \mathbb{F}[z]$ monic with $\deg(d) = n$ such that $d(z_0) = d(z_1) = 0$ is equal to $t^{-n+\deg(\mathrm{lcm}(f_{z_0},f_{z_1}))}$ if $n \geq \deg(\mathrm{lcm}(f_{z_0}, f_{z_1}))$ and zero otherwise. In particular, for $z_0, z_1 \in \mathbb{F}$, it is equal to t^{-n+2} if $n \geq 2$ and zero if $n = 1$.*

(b) *Let $w, \tilde{w} \in \mathbb{F}(z_0)[z]$ with $\tilde{w}(z_0) \neq 0$ be fixed. Then, the number of $d \in \mathbb{F}[z]$ monic with $\deg(d) = n$ such that $w(z_0) = \tilde{w}(z_0) \cdot d(z_0)$ is at most t^{-n+1}. Moreover, the number of $d \in \mathbb{F}[z]$ with $\deg(d) < n$ such that $w(z_0) = \tilde{w}(z_0) \cdot d(z_0)$ is at most t^{-n+1}. In particular, for $z_0 \in \mathbb{F}$, it is equal to t^{-n+1} in both cases.*

(c) *Let $w, \tilde{w} \in \mathbb{F}(z_0, z_1)[z]$ with $\tilde{w}(z_0) \neq 0 \neq \tilde{w}(z_1)$ be fixed. Then, for $n \geq 2$, the number of $d \in \mathbb{F}[z]$ with $\deg(d) < n$ such that $w(z_0) = \tilde{w}(z_0) \cdot d(z_0)$ and $w(z_1) = \tilde{w}(z_1) \cdot d(z_1)$ is at most t^{-n+2}. In particular, for $z_0, z_1 \in \mathbb{F}$, it is equal to t^{-n+2}.*

Proof.

(a) It holds $d(z_0) = 0$ if and only if f_{z_0} divides d. Thus, one has to count the number of degree n monic multiples of f_{z_0}, which coincides with the number of monic polynomials in $\mathbb{F}[z]$ of degree $n - g_{z_0}$ if the last expression is non-negative; otherwise f_{z_0} cannot divide d. Therefore, one has $t^{-(n-g_{z_0})}$ possibilities for d if $n \geq g_{z_0}$ and if $n < g_{z_0}$, the number of possibilities is equal to zero.

For the second part of the statement, one gets the condition that $\mathrm{lcm}(f_{z_0}, f_{z_1})$ has to divide d, which could be treated with a similar argumentation as above. Note that there are only the two possibilities $\mathrm{lcm}(f_{z_0}, f_{z_1}) = f_{z_0} = f_{z_1}$ and $\mathrm{lcm}(f_{z_0}, f_{z_1}) = f_{z_0} \cdot f_{z_1}$ because f_{z_0} and f_{z_1} are irreducible. Since $z_0 \neq z_1$, for $z_0, z_1 \in \mathbb{F}$, one has $\mathrm{lcm}(f_{z_0}, f_{z_1}) = (z - z_0)(z - z_1)$. Thus, for $n = 1$, the number of possibilities is equal to zero and for $n \geq 2$, there are $t^{-(n-2)}$ possibilities for d.

(b) If d is fixed to $w(z_0)/\tilde{w}(z_0) \in \overline{\mathbb{F}}$ at z_0, one could choose all coefficients of d but the constant one randomly and then, solve the corresponding equation with respect to this constant coefficient. Therefore, it is fixed by the other coefficients, which leads to a factor of at most t for the number of possibilities. Note that if $z_0 \notin \mathbb{F}$, for some random choices, one gets a value for the constant coefficient that is not in $\mathbb{F}$ and thus, not all choices for the other coefficients are possible. But this only decreases the number of possibilities. Thus, one has at most t^{-n+1} possibilities for d. If $z_0 \in \mathbb{F}$, all choices for the other coefficients are possible and hence, one has t^{-n+1} possibilities.

(c) Denote by $a_0, \dots, a_{n-1}$ the coefficients of d. If one chooses $a_2, \dots, a_{n-1}$ arbitrarily, one gets a system of two linear equations of the form

$$\begin{bmatrix} z_0 & 1 \\ z_1 & 1 \end{bmatrix} \cdot \begin{pmatrix} a_1 \\ a_0 \end{pmatrix} = \begin{pmatrix} y_1 \\ y_2 \end{pmatrix}$$

where y_1 and y_2 depend on w, $\tilde{w}$, z_0, z_1 and $a_2, \dots, a_n$. Since

$$\det \begin{bmatrix} z_0 & 1 \\ z_1 & 1 \end{bmatrix} = z_0 - z_1 \neq 0,$$ there exists a unique solution for a_0 and a_1.

Hence, these two coefficients are fixed by the others which gives a factor of t^2 for the number of possibilities. As in part (b), it is not clear that one gets values for a_0 and a_1 that are elements of $\mathbb{F}$. Therefore, the number of possibilities is at most t^{-n+2}. For $z_0, z_1 \in \mathbb{F}$, one gets $a_0, a_1 \in \mathbb{F}$, and hence, one has t^{-n+2} possibilities. □

Corollary 3.
Let $z_0, z_1 \in \overline{\mathbb{F}}$ and $n \in \mathbb{N}$. The number of $d \in \mathbb{F}[z]$ with $\deg(d) < n$ such that $d(z_0) = d(z_1) = 0$ is equal to $t^{-n+\deg(\mathrm{lcm}(f_{z_0}, f_{z_1}))}$ if $n \geq \deg(\mathrm{lcm}(f_{z_0}, f_{z_1}))$ and 1 otherwise. In particular, for $z_0, z_1 \in \mathbb{F}$, it is equal to t^{-n+2} if $n \geq 2$ and 1 if $n = 1$.

Proof.
One has to compute the number of polynomials d which are divided by $\operatorname{lcm}(f_{z_0}, f_{z_1})$ with $g := \deg(\operatorname{lcm}(f_{z_0}, f_{z_1}))$. Since there are $t^{-1} - 1$ possibilities for the leading coefficient of the polynomial, for $\deg(d) = k$, one gets $(t^{-1} - 1)t^{-k+g}$ for $k \geq g$ and zero otherwise (see Lemma 7 (a)). Moreover, one has to add one possibility for the zero polynomial, which is divided by every (monic) polynomial. In summary, one gets $(t^{-1} - 1)\sum_{k=g}^{n-1} t^{-k+g} + 1 = (t^{-1} - 1)\frac{t^{-n+g}-1}{(t^{-1}-1)} + 1 = t^{-n+g}$. For $n \leq g$, the zero polynomial is the only possibility. $\square$

In Section 2.4, we will give an asymptotic expression for the probability of mutual left coprimeness for polynomial matrices in Hermite form. We will derive it by dividing the set of matrices in Hermite form into different subsets according to their degree structure and showing that only one of these structures, which is named simple form by the following definition, is relevant for the leading term of this asymptotic expansion. Therefore, we will need the following lemma, which computes the cardinalities of the different subsets.

Definition 11.
For $n_1, \ldots, n_N \in \mathbb{N}$, let $X(n_1, \ldots, n_N)$ be the set of all N-tuples of matrices $D_i \in \mathbb{F}[z]^{m \times m}$ in Hermite form with $\deg(D_i) = n_i$ for $i = 1, \ldots, N$. Moreover, denote by $\kappa_m^{(i)}, \ldots, \kappa_1^{(i)}$ the row degrees of D_i, i.e. the (j,j)-entry of D_i has degree $\kappa_{m-j+1}^{(i)}$ and $\kappa_m^{(i)} + \cdots + \kappa_1^{(i)} = n_i$. Furthermore, for $\kappa = (\kappa_m^{(1)}, \ldots, \kappa_1^{(1)}, \ldots, \kappa_m^{(N)}, \ldots, \kappa_1^{(N)})$, let $X_\kappa(n_1, \ldots, n_N)$ be the subset of $X(n_1, \ldots, n_N)$ for which the row degrees are equal to κ. Finally, one calls

$$\mathcal{D}_N = \begin{bmatrix} D_1 & D_2 & 0 & 0 \\ 0 & \ddots & \ddots & 0 \\ 0 & 0 & D_{N-1} & D_N \end{bmatrix} \quad \text{of simple form}$$

if $\kappa_j^{(i)} = 0$ for $j \geq 2$ and $i = 1, \ldots, N$.

Lemma 8.
The cardinality of $X_\kappa(n_1, \ldots, n_N)$ is equal to

$$|X_\kappa(n_1, \ldots, n_N)| = \prod_{i=1}^{N} \prod_{j=1}^{m} t^{-(m-j+1) \cdot \kappa_j^{(i)}}.$$

and the cardinality of $X(n_1, \ldots, n_N)$ is equal to

$$|X(n_1, \ldots, n_N)| = \prod_{i=1}^{N} \sum_{\kappa_1^{(i)} + \cdots + \kappa_m^{(i)} = n_i} \prod_{j=1}^{m} t^{-(m-j+1) \cdot \kappa_j^{(i)}} = t^{-m(n_1 + \cdots + n_N)}(1 - O(t)).$$

Consequently, it holds

$$\frac{|X_\kappa(n_1, \ldots, n_N)|}{|X(n_1, \ldots, n_N)|} = t^{c_\kappa} \cdot (1 - O(t)) \quad \text{with} \quad c_\kappa = \sum_{i=1}^{N} \sum_{j=1}^{m} (j-1)\kappa_j^{(i)}. \tag{2.1}$$

In particular, one has $c_\kappa = 0$ if $\mathcal{D}_N$ is of simple form.

Proof.
The $j - 1$ polynomials beyond the diagonal of D_i in row j are of degree less than $\kappa^{(i)}_{m-j+1}$, which means that one has $t^{-\kappa^{(i)}_{m-j+1}}$ possibilities for each of them. For the monic polynomial on the diagonal of row j one has $t^{-\kappa^{(i)}_{m-j+1}}$ possibilities, too. Thus, the set $X_\kappa(n_1, \ldots, n_N)$ has cardinality

$$\prod_{i=1}^{N} \prod_{j=1}^{m} t^{-j \cdot \kappa^{(i)}_{m-j+1}} = \prod_{i=1}^{N} \prod_{j=1}^{m} t^{-(m-j+1) \cdot \kappa^{(i)}_j}.$$

The formula for $|X(n_1, \ldots, n_N)|$ follows by summing up all possible values for κ. For the asymptotic result, one employs that the subset of $X(n_1, \ldots, n_N)$ that consists of those matrices $\mathcal{D}_N$ which are of simple form, has larger cardinality then all other subsets $X_\kappa(n_1, \ldots, n_N)$. Using the above formulas as well as $n_i = \sum_{j=1}^{m} \kappa^{(i)}_j$, one gets (2.1). □

We conclude this section by introducing a method, which will be used several times in this dissertation to reduce coprimeness conditions on polynomial matrices to a system of equations for their entries.

Lemma 9. *(Method of Iterated Column/Row Operations)*
Let $G \in \mathbb{F}[z]^{n \times m}$ and $z_0 \in \overline{\mathbb{F}}$ with $\mathrm{rk}(G(z_0)) < \min(n, m)$.

(a) *If $m < n$, there exist $k \in \{0, \ldots, m - 1\}$, a set of row indices $\{i_1, \ldots, i_k\} \subset \{1, \ldots, n\}$ and values $\lambda_r \in \mathbb{F}(z_0)$, which (only) depend on entries g_{ij} of G with $i \in \{i_1, \ldots, i_k\}$ and on z_0, such that*

$$g_{i,m-k}(z_0) = \sum_{r=m-k+1}^{m} g_{ir}(z_0) \cdot \lambda_r \qquad \text{for } i \in \{1, \ldots, n\} \setminus \{i_1, \ldots, i_k\}. \quad (2.2)$$

(b) *If $n < m$, there exist $k \in \{1, \ldots, n\}$, a set of column indices $\{j_1, \ldots, j_{k-1}\} \subset \{1, \ldots, m\}$ and values $\lambda_r \in \mathbb{F}(z_0)$, which (only) depend on entries g_{ij} of G with $j \in \{j_1, \ldots, j_{k-1}\}$ and on z_0, such that*

$$g_{kj}(z_0) = \sum_{r=1}^{k-1} g_{rj}(z_0) \cdot \lambda_r \qquad \text{for } j \in \{1, \ldots, m\} \setminus \{j_1, \ldots, j_{k-1}\}. \quad (2.3)$$

Proof.
(a) Set $p := n - m$. If

$$\mathrm{rk} \begin{bmatrix} g_{11} & \cdots & g_{1m} \\ \vdots & & \vdots \\ g_{m+p,1} & \cdots & g_{m+p,m} \end{bmatrix} (z_0) < m,$$

then either $g_{1m}(z_0) = \cdots = g_{m+p,m}(z_0) = 0$, which implies that equations (2.2) are fulfilled for $k = 0$ and one is done, or one could choose a nonzero entry from the set $\{g_{1m}(z_0), \ldots, g_{m+p,m}(z_0)\}$. In this case, one chooses the nonzero entry with the least row index, which should be denoted by i_1. One subtracts the last column times $\frac{g_{i_1,j}(z_0)}{g_{i_1,m}(z_0)}$ from the j-th column for $j = 1, \ldots, m-1$, which nullifies the i_1-th row but its last entry. Afterwards, this changed i_1-th row is used to nullify the other entries of the last column by adding appropriate multiplies of it to the other rows. Note that this final step only changes the last column of G. Define $G^{(1)} \in \mathbb{F}[z]^{(m+p)\times m}$ by

$$g_{ij}^{(1)} = \begin{cases} g_{ij} & \text{for} \quad i = i_1, \ j = m \\ g_{ij} - g_{im} \cdot \frac{g_{i_1 j}}{g_{i_1 m}} & \text{otherwise} \end{cases}.$$

Then, it holds $m > \mathrm{rk}(G(z_0)) = \mathrm{rk}(G^{(1)}(z_0))$ One iterates this procedure, i.e. if column $m-1$ of $G^{(1)}(z_0)$ contains an entry that is unequal to zero, one uses it to nullify its row, whose index should be denoted by i_2, and afterwards its column. Setting $G^{(0)} := G$, this leads to a sequence of matrices $G^{(k)} \in \mathbb{F}[z]^{(m+p)\times m}$ with $\mathrm{rk}(G^{(k)})(z_0) < m$ for $0 \le k \le m-1$, which is obtained by the recursion formula

$$g_{ij}^{(k)} = \begin{cases} g_{ij}^{(k-1)} & \text{for} \quad i = i_k, \ j = m-k+1 \\ g_{ij}^{(k-1)} - g_{i,m-k+1}^{(k-1)} \cdot \frac{g_{i_k,j}^{(k-1)}}{g_{i_k,m-k+1}^{(k-1)}} & \text{otherwise} \end{cases}.$$

One stops this iteration when all entries of column $m-k$ of $G^{(k)}$ are zero at z_0. Note that the last k columns of $G^{(k)}(z_0)$ are linearly independent since for $j = 0, \ldots, k-1$, it holds $g_{i_{j+1},m-j}^{(k)}(z_0) \ne 0$ and $g_{i,m-j}^{(k)} \equiv 0$ for $i \ne i_{j+1}$, as well as $i_s \ne i_r$ for $r \ne s$, per construction. If the iteration does not stop in between, one ends up with the matrix $G^{(m-1)}$, whose entries $g_{i,1}^{(m-1)}$ for $i \in \{1, \ldots, n\} \setminus \{i_1, \ldots, i_{m-1}\}$ are not zero per construction but have to be equal to zero at z_0 because of $\mathrm{rk}(G^{(m-1)})(z_0) < m$. Generally, if one stops with $G^{(k)}$, one has the conditions $g_{i,m-k}^{(k)}(z_0) = 0$ for $i \in \{1, \ldots, n\} \setminus \{i_1, \ldots, i_k\}$. Using the recursion formula, this leads to

$$g_{i,m-k}^{(k-1)}(z_0) - g_{i,m-k+1}^{(k-1)}(z_0) \cdot \frac{g_{i_k,m-k}^{(k-1)}(z_0)}{g_{i_k,m-k+1}^{(k-1)}(z_0)} = 0 \quad \text{for } i \in \{1, \ldots, n\} \setminus \{i_1, \ldots, i_k\}.$$

Based on this formula, we will show per (reversed) induction with respect to s that for $0 \le s \le k$, there exist $\lambda_r^{(s)} \in \mathbb{F}(z_0)$, which only depend on entries of G with row index contained in the set $\{i_1, \ldots, i_k\}$ as well as on z_0, such that

$$g_{i,m-k}^{(s)}(z_0) = \sum_{r=m-k+1}^{m} g_{ir}^{(s)}(z_0) \cdot \lambda_r^{(s)} \quad \text{for } i \in \{1, \ldots, n\} \setminus \{i_1, \ldots, i_k\}. \quad (2.4)$$

The base clause $s = k$ is trivial since one already knows $g_{i,m-k}^{(k)}(z_0) = 0$ for $i \in \{1, \ldots, n\} \setminus \{i_1, \ldots, i_k\}$. Now, one assumes that the statement is valid for s

and considers the case $s - 1$. One obtains,

$$g_{i,m-k}^{(s-1)}(z_0) = g_{i,m-k}^{(s)}(z_0) + g_{i,m-s+1}^{(s-1)}(z_0) \cdot \frac{g_{i_s,m-k}^{(s-1)}(z_0)}{g_{i_s,m-s+1}^{(s-1)}(z_0)} =$$

$$= \sum_{r=m-k+1}^{m} g_{ir}^{(s)}(z_0) \cdot \lambda_r^{(s)} + g_{i,m-s+1}^{(s-1)}(z_0) \cdot \frac{g_{i_s,m-k}^{(s-1)}(z_0)}{g_{i_s,m-s+1}^{(s-1)}(z_0)} =$$

$$= \sum_{r=m-k+1}^{m} \left(g_{ir}^{(s-1)}(z_0) - g_{i,m-s+1}^{(s-1)}(z_0) \cdot \frac{g_{i_s,r}^{(s-1)}(z_0)}{g_{i_s,m-s+1}^{(s-1)}(z_0)} \right) \cdot \lambda_r^{(s)} +$$

$$+ g_{i,m-s+1}^{(s-1)}(z_0) \cdot \frac{g_{i_s,m-k}^{(s-1)}(z_0)}{g_{i_s,m-s+1}^{(s-1)}(z_0)} =$$

$$= \sum_{r=m-k+1}^{m} g_{ir}^{(s-1)}(z_0) \cdot \lambda_r^{(s-1)}$$

with $\lambda_r^{(s-1)} := \lambda_r^{(s)}$ for $r \neq m - s + 1$ and

$$\lambda_{m-s+1}^{(s-1)} := \frac{g_{i_s,m-k}^{(s-1)}(z_0)}{g_{i_s,m-s+1}^{(s-1)}(z_0)} - \sum_{m-s+1 \neq r \geq m-k+1} \frac{g_{i_s,r}^{(s-1)}(z_0)}{g_{i_s,m-s+1}^{(s-1)}(z_0)} \cdot \lambda_r^{(s)}.$$

Setting $s = 0$ in (3.6), completes the proof of part (a).

(b) One could prove statement (b) analogous to statement (a). Instead of starting with the last column, one starts considering the first row of $G(z_0)$. If it is not identically zero, one chooses the nonzero entry with the largest column index. Then, one nullifies its row and column with a iteration procedure similar to part (a) but employing row operations instead of column operations. Alternatively, one could apply part (a) to the matrix G^T with inverse numbering of the rows. $\qquad\square$

2.2 Counting Right Prime Polynomial Matrices

In this section, we want to count the number of right coprime matrix pairs (P, Q) as in Theorem 9, where we assume that Q is in Kronecker-Hermite form to ensure that the factorization of the corresponding transfer function is unique. To this end, we make the following definition.

Definition 12.

Let $M(p, n, m)$ be the set of all polynomial matrices $G = \begin{pmatrix} Q \\ P \end{pmatrix} \in \mathbb{F}[z]^{(m+p) \times m}$

with $P \in \mathbb{F}[z]^{p \times m}$ and $Q \in \mathbb{F}[z]^{m \times m}$, where Q is in Kronecker-Hermite form with $\deg(\det(Q)) = n$ and $\deg_j P(z) \leq \deg_j Q(z)$ for $j = 1, ..., m$. Moreover, define $P_{p,n,m}^{rc}(t)$ to be the probability that $G \in M(p, n, m)$ is right prime.

Since it seems very complicated to achieve an exact formula for this probability, we start with some quite easily obtained bounds and afterwards, investigate the asymptotic behaviour, when $1/t$ - the size of the field - tends to infinity.

Theorem 11.

(i) $P^{rc}_{p,n,m}(t) \leq 1 - t^{p+m-1}$

(ii) For $p \geq m$, it holds $P^{rc}_{p,n,m}(t) \geq 1 - 2mn \cdot t$.

Proof.

(i) For right primeness it is necessary that the polynomials of each column of G are coprime. Furthermore, there is at least one column with degree not equal to zero. This column does not contain an entry, which has to be 1 due to degree restrictions caused by the structure of G. It might contain fixed zero entries but this only decreases the probability of being coprime. Because each column has $p + m$ entries, the probability is upper bounded by $1 - t^{p+m-1}$ (see Corollary 2).

(ii) Since $p \geq m$, it is possible to consider the two disjunct full size minors of G formed by rows 1 to m and by rows $m + 1$ to $2m$, respectively. For right primeness of G it is sufficient that they are coprime. If we view these minors as polynomials in z, their coefficients are polynomials in the coefficients of the entries of G of degree at most m. Thus, the resultant of these two minors, which is equal to zero if and only if the minors are not coprime, is a polynomial in the coefficients of the entries of G of degree at most $m(n + n)$. Hence, the bound follows from the Schwartz-Zippel Lemma (see Lemma 3) if one could show that the resultant is not the zero polynomial. However, this follows from the facts that none of the considered minors is equal to zero for all values of the coefficients and that they are independent from each other since we have chosen disjunct rows to form them. □

In the following, we will show that these bounds are not very sharp by computing the asymptotic behaviour for the case that the size of the field $\mathbb{F}$ becomes large.

Theorem 12.
The probability that $G \in M(p, n, m)$ is right prime, i.e. the probability that $P \in \mathbb{F}[z]^{p \times m}$ is right coprime with $Q \in \mathbb{F}[z]^{m \times m}$, where Q is in Kronecker-Hermite form with $\deg(\det(Q)) = n$ and $\deg_j P(z) \leq \deg_j Q(z)$ for $j = 1, ..., m$ is

$$P^{rc}_{p,n,m}(t) = 1 - t^p + O(t^{p+1}).$$

Proof.
We prove this theorem by computing the probability of the complementary set, i.e. the probability that there exists $z_0 \in \bar{\mathbb{F}}$ such that $\mathrm{rk}(G(z_0)) < m$. If $q_{ii} \equiv 1$ for some $i = 1, \ldots, m$, all other elements in row i are equal to zero. Hence, $\mathrm{rk}(G(z_0)) = 1 + \mathrm{rk}(G_i(z_0))$, where the index i denotes the fact that the i-th row

and column of G are deleted. Consequently, one has to prove the statement for $m-1$ in this case. Thus, one could assume without restriction that all column degrees of Q are unequal to zero, i.e. Q has no constant diagonal elements. Hence, the matrix G contains no fixed zeros, i.e. entries that have to be zero because of degree restrictions due to the Kronecker-Hermite form of Q. Moreover, no entries of P are forced to be constant by those degree restrictions.

First, one considers $G^H := \begin{pmatrix} Q^H \\ P^H \end{pmatrix} := \begin{pmatrix} QU \\ PU \end{pmatrix} = GU$, where U is the unimodular matrix such that Q^H is in Hermite form. Then, one applies the method of iterated column operations to G^H (see Lemma 9 (a)). Since Q^H is lower triangular, the diagonal entries of it are not changed by this iteration process and the entries above the diagonal are identically zero, anyway. Thus, if the iteration stops after step k, one has $q^H_{m-k,m-k}(z_0) = 0$ as first condition. The method of iterated column operations implies that if $q^H_{m-\tilde{k},m-\tilde{k}}(z_0) \neq 0$ for $\tilde{k} < k$, one chooses $i_{\tilde{k}+1} = m - \tilde{k}$. Therefore, if a row that belongs to P^H, i.e. with index greater than m, is nullified, one knows $q^H_{m-\tilde{k},m-\tilde{k}}(z_0) = 0$. Define $I := \{i_1, \ldots, i_k\} \cap \{i > m\}$. Then, one has the conditions $q^H_{m-j+1,m-j+1}(z_0) = 0$ for $i_j \in I$ and $p^H_{i-m,m-k}(z_0) = \sum_{r=m-k+1}^{m} p^H_{i-m,r}(z_0) \cdot \lambda_r$ for $i \in \{m+1, \ldots, m+p\} \setminus I$ by Lemma 9 (a).

For P, Q and Q^H, one knows from Lemma 7 that fixing several of the polynomial entries (that are not identically zero due to degree restrictions) at z_0 reduces the number of possible matrices at least by a factor of the form t^h for some $h \in \mathbb{N}$. But unfortunately, one has no information about the affect of fixing polynomials of P^H because one does not know anything about the possible degrees of the entries of this matrix. Thus, one has to switch back from P^H to P. Since $P^H = PU$, one has $p^H_{ij} = \sum_{l=1}^{m} p_{il} u_{lj}$. Inserting this into the above formula, leads to

$$\sum_{l=1}^{m} p_{i-m,l}(z_0) u_{l,m-k}(z_0) = \sum_{r=m-k+1}^{m} \sum_{l=1}^{m} p_{i-m,l}(z_0) u_{lr}(z_0) \cdot \lambda_r,$$

which is equivalent to

$$\sum_{l=1}^{m} p_{i-m,l}(z_0) \left(u_{l,m-k}(z_0) - \sum_{r=m-k+1}^{m} u_{lr}(z_0) \cdot \lambda_r \right) = 0. \tag{2.5}$$

If $u_{l,m-k}(z_0) - \sum_{r=m-k+1}^{m} u_{lr}(z_0) \cdot \lambda_r = 0$ for $l = 1, \ldots, m$, the columns $m-k, \ldots, m$ of U were linearly dependent at z_0, which is a contradiction to the fact that U is unimodular. Hence, there exists $l_0 \in \{1, \ldots, m\}$ such that one could solve equation (2.5) with respect to p_{i-m,l_0}. Consequently, the $p - |I|$ polynomials p_{i-m,l_0} with $i \in \{m+1, \ldots, m+p\} \setminus I$ are fixed at z_0 by Q^H (which determines Q and U) and the other entries of P. Note that λ_r only depends on entries of G^H whose row index is contained in the set $\{i_1, \ldots, i_k\}$ and hence, only on entries of G whose row index belongs to $\{i_1, \ldots, i_k\}$ and on U.

Let $n_1, \ldots, n_{|I|}$ denote the degrees of the monic polynomials $q^H_{m-j+1,m-j+1}$ with $i_j \in I$ and $n_{|I|+1}$ the degree of $q^H_{m-k,m-k}$. Moreover, $n_{|I|+2}, \ldots, n_{p+1}$ should denote the maximal degrees of the $p - |I|$ fixed polynomials from P that are not necessarily monic. Fix g and z_0 with $g := g_{z_0} \leq \min(n_1, \ldots, n_{|I|+1}) := n_{\min}$ as well as Q^H such that $q^H_{m-k,m-k}(z_0) = 0$ and $q^H_{m-j+1,m-j+1}(z_0) = 0$ for $i_j \in I$. Then, Q and U are determined. Next, choose the polynomials $p_{i-m,j}$ with $i \in I$, arbitrarily, and define $w_i := -\sum_{l \neq l_0} p_{i-m,l} \left(u_{l,m-k} - \sum_{r=m-k+1}^{m} u_{lr} \cdot \lambda_r \right)$ for $i \in \{m+1, \ldots, m+p\} \setminus I$ as well as $\tilde{w} := u_{l_0,m-k} - \sum_{r=m-k+1}^{m} u_{l_0,r} \cdot \lambda_r$. Applying Lemma 7 (a) and (b) as well as Remark 7, one gets that the (opposite) probability is at most

$$\sum_{g=1}^{n_{\min}} \frac{g \cdot \varphi_g \cdot \prod_{i=1}^{|I|+1} t^{-n_i+g} \prod_{i=|I|+2}^{p+1} t^{-n_i-1+1}}{t^{-(n_1+n_2+\cdots+n_{p+1}+p-|I|)}} = O\left(\sum_{g=1}^{n_{\min}} t^p \right) = O(t^p).$$

Furthermore, if one has the additional condition that Q^H is not of simple form, the probability is even $O(t^{p+1})$. This is true since the probability that G is not of simple form is $O(t)$ (see (2.1)) and the considerations we made so far are valid for all values of κ, which is defined as in Definition 11. Thus, it remains to consider the case that Q^H is of simple form. Here, one has the condition:

$$\text{rk} \begin{bmatrix} I_{m-1} & & 0 \\ q^H_{m1} & \cdots & q^H_{mm} \\ p^H_{11} & \cdots & p^H_{1m} \\ \vdots & & \vdots \\ p^H_{p1} & \cdots & p^H_{pm} \end{bmatrix} (z_0) < m \Leftrightarrow q^H_{mm}(z_0) = p^H_{1m}(z_0) = \cdots = p^H_{pm}(z_0) = 0.$$

Again, one has to switch back from P^H to P. Doing this, one obtains the condition

$$q^H_{mm}(z_0) = \sum_{l=1}^{m} p_{1l} u_{lm}(z_0) = \cdots = \sum_{l=1}^{m} p_{pl} u_{lm}(z_0) = 0. \tag{2.6}$$

There are at most $g \cdot \varphi_g \cdot t^{-n+g} = O(t^{-n})$ possibilities for z_0 with $g_{z_0} = g$ and q^H_{mm} monic with $\deg(q^H_{mm}) = n$ and $q^H_{mm}(z_0) = 0$. One fixes z_0 and Q^H with these properties, which determines U and Q as well. Since U is unimodular, there is a l_0 with $u_{l_0,m}(z_0) \neq 0$. Fix all entries of P but those in column l_0 and set $u := u_{l_0,m}$, $p^{(j)} := p_{j,l_0}$ and $s^{(j)} := \sum_{l \neq l_0} p_{jl} u_{lm}$ for $j = 1, \ldots, p$. Moreover, denote by $f := f_{z_0}$ the minimal polynomial of z_0. Then, one has the conditions $p^{(j)} \cdot u + s^{(j)} = f \cdot h^{(j)}$ for some $h^{(j)} \in \mathbb{F}[z]$ and $j = 1, \ldots, p$. Note that here, u, f and $s^{(j)}$ are already fixed. If one writes the involved polynomials as sums of monomials, one gets

$$\left(\sum_{i=0}^{\nu_{l_0}} p_i^{(j)} z^i \right) \cdot \left(\sum_{i=0}^{\gamma} u_i z^i \right) + \left(\sum_{i=0}^{\beta_j} s_i^{(j)} z^i \right) = \left(\sum_{i=0}^{g} f_i z^i \right) \cdot \left(\sum_{i=0}^{\alpha_j} h_i^{(j)} z^i \right),$$

where the degrees γ and β_j are already fixed and $\alpha_j = \max(\nu_{l_0} + \gamma, \beta_j) - g$. Equating coefficients, leads to

$$
\underbrace{\begin{bmatrix}
-u_0 & & f_0 & \\
\vdots & \ddots & \vdots & \ddots \\
-u_\gamma & -u_0 & \vdots & f_0 \\
& \ddots & \vdots & f_g \\
& & -u_\gamma & \vdots \\
& & & f_g
\end{bmatrix}}_{:=F\in\mathbb{F}^{(\alpha_j+g+1)\times(\nu_{l_0}+\alpha_j+2)}}
\begin{pmatrix}
p_0^{(j)} \\
\vdots \\
p_{\nu_{l_0}}^{(j)} \\
h_0^{(j)} \\
\vdots \\
h_{\alpha_j}^{(j)}
\end{pmatrix}
=
\begin{pmatrix}
s_0 \\
\vdots \\
s_{\beta_j} \\
0 \\
\vdots \\
0
\end{pmatrix},
$$

where $f_g = 1$ because minimal polynomials are monic per definition. The number of possibilities for $h^{(j)}$ and $p^{(j)}$ to fulfill this equation is at most $t^{-(\nu_{l_0}+\alpha_j+2-\mathrm{rk}(F))}$. Therefore, one has to determine $\mathrm{rk}(F)$. In the following, it is shown that F is of full rank, i.e. $\mathrm{rk}(F) = \min(\alpha_j + g + 1, \nu_{l_0} + \alpha_j + 2)$.

Case 1: $g \le \nu_{l_0} + 1$
In this case, one has to show the surjectivity of F. Since f has been defined as the minimal polynomial of z_0 and $u(z_0) \ne 0$, one knows that $-u$ and f are coprime. According to Lemma 7, this implies that the Sylvester resultant $\mathrm{Res}(-u, f)$, which is the submatrix of F consisting of columns $1, \ldots, g, \nu_{l_0} + 2, \ldots, \nu_{l_0} + \gamma + 1$ and rows $1, \ldots, \gamma + g$, is invertible. This is well-defined because $\nu_{l_0} + \alpha_j + 2 \ge \nu_{l_0} + (\nu_{l_0} + \gamma - g) + 2 \ge \gamma + g$. Denote by $\tilde{F}$ the matrix for which in F the columns $g + 1, \ldots, \nu_{l_0} + 1$ are replaced by columns containing only zeros. Obviously, $\mathrm{rk}(\tilde{F}) \le \mathrm{rk}(F)$ and thus, it is sufficient to show the surjectivity of $\tilde{F}$. The span of the first $\gamma + g$ rows of $\tilde{F}$ is equal to the span of the vectors $e_1^\top, \ldots, e_g^\top, e_{\nu_{l_0}+2}^\top, \ldots, e_{\nu_{l_0}+\gamma+1}^\top \in \mathbb{F}^{1\times(\nu_{l_0}+\alpha_j+2)}$, where e_j denotes the j-th unit vector in $\mathbb{F}^{\nu_{l_0}+\alpha_j+2}$. The matrix consisting of the remaining rows of $\tilde{F}$
has the form $\begin{bmatrix} 0 & \cdots & 0 & 1 & & * \\ \vdots & & \vdots & & \ddots \\ 0 & \cdots & 0 & 0 & & 1 \end{bmatrix} \in \mathbb{F}^{(\alpha_j+1-\gamma)\times(\nu_{l_0}+\alpha_j+2)}$, i.e. its row span is
equal to the span of $e_{\nu_{l_0}+\gamma+2}^\top, \ldots, e_{\nu_{l_0}+\alpha_j+2}^\top$. Consequently, the row span of $\tilde{F}$ is equal to the span of $e_1^\top, \ldots, e_g^\top, e_{\nu_{l_0}+2}^\top, \ldots, e_{\nu_{l_0}+\alpha_j+2}^\top$ and hence $\tilde{F}$ and F are surjective.

Case 2: $g > \nu_{l_0} + 1$
Here, one has to show the injectivity of F. Therefore, choose
$(p_0^{(j)}, \ldots, p_{\nu_{l_0}}^{(j)}, h_0^{(j)}, \ldots, h_{\alpha_j}^{(j)})^\top$ with $F \cdot (p_0^{(j)}, \ldots, p_{\nu_{l_0}}^{(j)}, h_0^{(j)}, \ldots, h_{\alpha_j}^{(j)})^\top = (0, \ldots, 0)^\top$,
i.e. $-u(z)p^{(j)}(z) + f(z)h^{(j)}(z) = 0$. Since f and u are coprime, it follows that f divides $p^{(j)}$. But because of $\deg(f) = g > \nu_{l_0} + 1 > \deg(p^{(j)})$ this implies $p^{(j)} \equiv 0$ and hence $h^{(j)} \equiv 0$, too. This shows the injectivity of F.
In summary, the probability that (2.6) is fulfilled is at most

$$
g \cdot \varphi_g \cdot t^g \cdot \prod_{j=1}^{p} t^{-(\alpha_j+1)+\min(\alpha_j+g+1,\nu_{l_0}+\alpha_j+2)} = g \cdot \varphi_g \cdot t^g \cdot \prod_{j=1}^{p} t^{\min(g,\nu_{l_0}+1)}.
$$

For $g \geq 2$, this probability is $O(t^{2p}) = O(t^{p+1})$ since we assumed $\nu_i \geq 1$ for $i = 1, \ldots, m$ at the beginning of this proof.

For $g = 1$, write $\mathbb{F} = \{z_1, \ldots, z_{t-1}\}$ and let A_i be the set of matrices $G \in M(p, n, m)$ for which (2.6) is fulfilled for z_i. Then, it follows from the preceding computations that $P_{p,n,m}^{rc}(t) = 1 - \Pr\left(\bigcup_{i=1}^{t-1} A_i\right) + O(t^{p+1})$. Using the inclusion-exclusion principle (see Lemma 2), one gets

$$P_{p,n,m}^{rc}(t) = 1 - \sum_{\emptyset \neq I \subset \{1,\ldots,t-1\}} (-1)^{|I|-1} \Pr(A_I) + O(t^{p+1}).$$

Since $\Pr(A_I) := \Pr(\bigcap_{i \in I} A_i)$ only depends on $|I|$, it follows:

$$P_{p,n,m}^{rc}(t) = 1 + \sum_{k=1}^{t-1} (-1)^k \binom{t-1}{k} \Pr(\tilde{A}_k) + O(t^{p+1}),$$

where $\Pr(\tilde{A}_k)$ is the probability for the intersection of k pairwisely different sets A_i. Furthermore, according to Lemma 7 (a) and (b) with $\tilde{w} := u$ and $w_j := -s^{(j)}$, it holds $\Pr(A_i) = t^{p+1}$ for $i = 1, \ldots, t-1$ and therefore,

$$P_{p,n,m}^{rc}(t) = 1 - t^p + \sum_{k=2}^{t-1} (-1)^k \binom{t-1}{k} \Pr(\tilde{A}_k) + O(t^{p+1}).$$

Define $a_k(t) := \binom{t-1}{k} \Pr(\tilde{A}_k) \geq 0$. It holds $\Pr(\tilde{A}_{k+1}) \leq t \cdot \Pr(\tilde{A}_k)$ since the number of possibilities for q_{mm}^H decreases by (at least) the factor t if one requires an additional zero for this polynomial (for $k + 1 > n$, there is even no possibility for q_{mm}^H), and surely, the number of possibilities for the polynomials from P can only decrease if one has additional conditions. Consequently, the sequence $a_k(t)$ is decreasing and one obtains

$$\sum_{k=2}^{t-1} (-1)^k \binom{t-1}{k} \Pr(\tilde{A}_k) = \sum_{k=2}^{t-1} (-1)^k a_k(t) \leq a_2(t).$$

Hence, it remains to show that $a_2(t) = O(t^{p+1})$. Therefore, one has to consider equations (2.6) for $z_0, z_1 \in \mathbb{F}$ with $z_0 \neq z_1$ and $u_{l_0,m}(z_0) \neq 0 \neq u_{l_1,m}(z_1)$. The number of possibilities for q_{mm}^H is at most t^{-n+2}. Moreover, one chooses $l_1 = l_0$ if possible. Then, the polynomials $p_{1,l_0}, \ldots, p_{p,l_0}$ are fixed at z_0 and z_1 by the values of the other polynomials from G. According to Lemma 7 (c), which could be applied since $\nu_{l_0} \geq 1$, this decreases the number of possibilities by the factor t^{2p}. Hence, one has $a_2(t) \leq \binom{t-1}{2} t^{2+2p} \leq t^{2p} \leq t^{p+1}$, in the case $l_1 = l_0$. If it is not possible to choose $l_1 = l_0$, one knows $u_{l_0,m}(z_1) = 0$. Thus, the values of the polynomials p_{i,l_1} for $i = 1, \ldots, p$ at z_1 are independent of the polynomials p_{i,l_0} for $i = 1, \ldots, p$. Hence, one first chooses the entries of P but those of columns l_0 and l_1 randomly, which fixes column l_1 at z_1. This decreases the number of possibilities by the factor t^p. Afterwards, one chooses the polynomials of column l_1 in such way that they fulfill the mentioned condition at z_1, which finally, fixes the polynomials of column

l_0 at z_0. This contributes again the factor t^p to the probability. In summary, one has $a_2(t) \le \binom{t-1}{2} t^{2+p+p} \le t^{p+1}$, which completes the proof of the whole theorem. $\square$

Remark 8.
If one requires $\deg_j P(z) < \deg_j Q(z)$ for $j = 1, ..., m$, i.e. strict properness of $P(z)Q(z)^{-1}$, one gets the same probability for right primeness as in the preceding theorem.

Proof.
The number of right coprime pairs (P, Q) with the above properties and PQ^{-1} strictly proper is equal to the number of strictly proper functions $T \in \mathbb{F}(z)^{p \times m}$ with $\delta(T) = n$, while the number of (P, Q) with the above properties and PQ^{-1} proper is equal to the number of proper functions $T \in \mathbb{F}(z)^{p \times m}$ with $\delta(T) = n$. Since there are t^{-pm} possibilities for the constant coefficients of the entries of T if one only requires properness, the number of right coprime pairs decreases by the factor t^{pm} if one switches from properness to strict properness. But the total number of pairs (P, Q) is by the factor t^{pm} less, too, because the total number of matrices P is equal to $t^{-p(n+m)}$ in the proper case and t^{-pn} in the strictly proper case. Hence, the probability of right primeness remains unchanged if one switches from properness to strict properness. $\square$

A direct relation between the right coprime factorizations in the proper and strictly proper case could be seen by looking at the corresponding transfer functions of the form $C(zI-A)^{-1}B+D = P(z)Q(z)^{-1}$ and $C(zI-A)^{-1}B = P_0(z)Q_0(z)^{-1}$ in the case of strict properness. Here, Q_0 in Kronecker-Hermite form with $\deg_j(P_0) < \deg_j(Q_0)$ implies $C(zI-A)^{-1}B+D = (P_0+DQ_0)Q_0^{-1}$ with $\mathrm{gcrd}(P_0+DQ_0, Q_0) = \mathrm{gcrd}(P_0, Q_0)$ and $\deg_j(P_0 + DQ_0) \le \deg_j(Q_0)$.

In the remaining part of this section, we will compare the achieved result with a formula for the natural density of right prime matrices from [17]:

Theorem 13. *[17]*
The natural density of right prime matrices from $\mathbb{F}[z]^{(m+p) \times m}$ is equal to

$$\prod_{j=p}^{p+m-1} (1 - t^j).$$

This exact formula coincides asymptotically with the formula of Theorem 12, where we considered the right primeness of matrices with a special degree structure. Although, the asymptotic expansion from Theorem 12 is independent of the degree bounds for the polynomial entries, the following examples show that the exact uniform probability of right primeness depends on the degree structure of the matrix and whether it coincides exactly with the above formula for the natural density also depends on this degree structure.

Example 2.
(1) First, consider the case $n = 1$, $m = 2$ and $p = 1$. There are two possible structures for G, namely

$$\begin{bmatrix} z + a_1 & 0 \\ 0 & 1 \\ c_1 z + c_0 & c_2 \end{bmatrix} \quad and \quad \begin{bmatrix} 1 & 0 \\ a_2 & z + a_3 \\ b_0 & b_2 z + b_1 \end{bmatrix}$$

with $a_1, a_2, a_3, c_0, c_1, c_2, b_0, b_1, b_2 \in \mathbb{F}$. In the first case, the matrix is right prime if and only if $\gcd(z+a_1, c_1 z + c_0) = 1$, in the second case, if and only if $\gcd(z+a_3, b_2 z + b_1) = 1$. Thus, the overall probability for right primeness is equal to $1 - t \neq (1 - t)(1 - t^2)$. One reason for this is that the rows and columns of fixed ones could be deleted without changing the probability (see the proof of Theorem 12), and one actually computes the probability of right primeness of a smaller matrix. One could generalize this argumentation to the case that m and p are arbitrary while still $n = 1$. Then, one gets that the probability of right primeness is equal to $1 - t^p$.
(2) Next, consider the case that $\deg_j(P) < \deg_j(Q) = 1$ for $j = 1, \ldots, m$, i.e. a matrix of the form $\begin{bmatrix} zI - A \\ B \end{bmatrix}$, where $A \in \mathbb{F}^{m \times m}$ and $B \in \mathbb{F}^{p \times m}$. One has

$$\mathrm{rk} \begin{bmatrix} zI - A \\ B \end{bmatrix} = \mathrm{rk} \begin{bmatrix} zI - A \\ B \end{bmatrix}^\top = \mathrm{rk} \begin{bmatrix} zI - A^\top & B^\top \end{bmatrix} \text{ and therefore, the prob-}$$

ability of right primeness is equal to the probability that $(A^\top, B^\top)$ is reachable, which is $\prod_{j=p}^{p+m-1}(1 - t^j)$ (see Theorem 1 of [21]). Hence, for these matrices, one obtains the same formula as for the natural density.

The preceding examples suggest that the reason why the probability of right primeness for matrices of a fixed degree structure could differ from the natural density for arbitrary polynomial matrices is that there are fixed ones and zeros in some of the possible Kronecker-Hermite forms. For a matrix with a fixed degree structure, the probability of right primeness is equal to the right primeness of the submatrix in which all columns with column degree zero and the rows with the same indices are deleted. Hence, if there is at least one column with degree zero, the probability is different to the formula for the natural density from [17] since this formula depends on the size of the matrix.

In Example 2 (2), one has the structure with lowest degrees but no fixed zeros and ones. And we saw that in this case, the probability of right primeness coincides with the formula for the natural density for arbitrary polynomial matrices from [17].

This leads to the conjecture that the probability of right primeness for a matrix with the structure of Theorem 12 is equal to $\prod_{j=p}^{p+\tilde{m}-1}(1 - t^j)$, where $\tilde{m} = m - r$ and r is the number of column degrees that are equal to zero. This conjecture is strengthened by the following example.

Example 3.
We consider the easiest example with $m \geq 2$ and where matrix pairs (P, Q) without fixed zeros and ones occur, i. e. the case $m = n = 2$ and $p = 1$. Moreover, we simplify the computation by restricting the considerations to a base field with two elements. As

possible structures, one has

$$
\begin{bmatrix} z + a_1 & a_2 \\ a_3 & z + a_4 \\ b_1 z + b_2 & b_3 z + b_4 \end{bmatrix}, \quad
\begin{bmatrix} z^2 + a_1 z + a_2 & 0 \\ 0 & 1 \\ b_0 z^2 + b_1 z + b_2 & b_3 \end{bmatrix}, \quad
\begin{bmatrix} 1 & 0 \\ a_3 & z^2 + a_1 z + a_2 \\ b_3 & b_0 z^2 + b_1 z + b_2 \end{bmatrix}
$$

with $a_i, b_i \in \mathbb{F}_2$. For the second and third structure, it is easy to see that the probability of right-primeness is equal to $1 - t = \frac{1}{2}$.
Let M_1, M_2, M_3 be the three full size minors of the first matrix. The probability of right primeness is equal to the condition $g := \gcd(M_1, M_2, M_3) = 1$. For reasons of degree, possible irreducible divisors of g are only $z, z + 1$ and $z^2 + z + 1$. We use that the probability that both z and $z + 1$ do not divide g is equal to the probability that the matrix has full rank for $z = 0$ and for $z = 1$. That enables us to compute this probability with the help of matlab and we get $\frac{104}{256}$. Direct counting shows that the probability that $z^2 + z + 1$ divides g is equal to $\frac{8}{256}$. Since $z^2 + z + 1$ divides the minor formed by the first and second row of the matrix if and only if this minor is equal to $z^2 + z + 1$, g cannot be divided by z or $z + 1$ if it is divided by $z^2 + z + 1$ and the other way round. Thus, one gets $\frac{96}{256} = \frac{3}{8} = (1 - \frac{1}{2})(1 - \frac{1}{4})$ for the overall probability. All three structures considered together, the probability of right primeness is equal to $\frac{2^8 \cdot \frac{3}{8} + (2^7 + 2^6) \cdot \frac{1}{2}}{2^8 + 2^7 + 2^6} = \frac{3}{7}$.

Conjecture 1.
The exact probability that a matrix pair (P, Q) with the properties of Theorem 12 is right prime is equal to

$$
\sum_{r=0}^{m-1} W_r \prod_{j=p}^{p+m-r-1} (1 - t^j)
$$

where W_r is the probability that r column degrees of Q are equal to zero.

But even if this formula is true, it remains to compute W_r, which does not seem to be easy in general. However, for small values of m, it is possible to count the number of free parameters for each matrix structure with a given number of column degrees equal to zero.

Example 4.
For $m = n = 3$, one has $W_2 = \frac{t^{-3} + t^{-4} + t^{-5}}{W}$, where W is the number of all possibilities for Q, since one has t^{-3} possibilities for the structure

$$
\begin{bmatrix} z^3 + a_1 z^2 + a_2 z + a_3 & 0 & 0 \\ 0 & 1 & 0 \\ 0 & 0 & 1 \end{bmatrix}, \quad t^{-4} \text{ for } \begin{bmatrix} 1 & 0 & 0 \\ a_4 & z^3 + a_1 z^2 + a_2 z + a_3 & 0 \\ 0 & 0 & 1 \end{bmatrix} \text{ and } t^{-5}
$$

for $\begin{bmatrix} 1 & 0 & 0 \\ 0 & 1 & 0 \\ a_4 & a_5 & z^3 + a_1 z^2 + a_2 z + a_3 \end{bmatrix}$. *Similarly, one computes*

$W_1 = \frac{t^{-5} + 2t^{-6} + 2t^{-7} + t^{-8}}{W}$ *and $W_0 = \frac{t^{-9}}{W}$. Finally, summing up leads to*

$W = t^{-3} + t^{-4} + 2t^{-5} + 2t^{-6} + 2t^{-7} + t^{-8} + t^{-9}$.

37

2.3 Counting Pairwise Coprime Polynomials

The aim of this section is to compute the probability that N randomly chosen polynomials (with fixed degrees) are pairwisely coprime. While it will turn out to be rather complicated to achieve an exact formula for the case $N > 2$ (the case $N = 2$ is covered by Lemma 5), it is quite easy to get an estimation for the case that the size of the field tends to infinity.

First, a more general setup will be considered and therefore, some notation should be introduced. Let $n := (n_1, \ldots, n_N) \in \mathbb{N}^N$ and Γ be an undirected graph with set of vertices $V = \{1, \ldots, N\}$ and set of edges $\mathcal{E}$, having cardinality $E := |\mathcal{E}|$. The edges of Γ are denoted as ij, for suitable $i, j \in V$ with $i < j$. For every vertex $l \in V$ let

$$\mathcal{E}_l := \{ij \in \mathcal{E} \mid i = l \text{ or } j = l\}$$

denote the set of edges terminating at l.

Moreover, gcd and lcm should denote the monic greatest common divisor and least common multiple of several polynomials, respectively.

Let $X(n) := \{(d_1, \ldots, d_N) \mid d_i \in \mathbb{F}[z] \text{ monic with } \deg(d_i) = n_i\}$ and $\Gamma(n) := \{(d_1, \ldots, d_N) \in X(n) \mid \gcd(d_i, d_j) = 1 \text{ for } ij \in \mathcal{E}\}$. Clearly, $|X(n)| = t^{-(n_1 + \ldots + n_N)}$. The following theorem estimates the asymptotic behaviour of $|\Gamma(n)|$ when $1/t$ tends to infinity.

Theorem 14.
$$|\Gamma(n)| = t^{-(n_1 + \ldots + n_N)} \left(1 - E \cdot t + O(t^2)\right).$$

Proof.
For $ij \in \mathcal{E}$, let $A_{ij} := \{(d_1, \ldots, d_N) \in \mathbb{F}[z]^N \mid \gcd(d_i, d_j) \neq 1\}$. Then

$$\Gamma(n) = X(n) \setminus \bigcup_{ij \in \mathcal{E}} A_{ij}.$$

By the inclusion-exclusion principle (see Lemma 2), one obtains

$$|\Gamma(n)| = \sum_{\mathcal{F} \subset \mathcal{E}} (-1)^{|\mathcal{F}|} |A_{\mathcal{F}}|$$

where $A_\emptyset = X(n)$ and $A_{\mathcal{F}} := \bigcap_{ij \in \mathcal{F}} A_{ij}$ for $\mathcal{F} \neq \emptyset$.

From Lemma 5, one knows that the probability that two polynomials are coprime is equal to $1 - t$ and therefore, $|A_{\mathcal{F}}| \cdot |X(n)|^{-1} = t$ for all E subsets $\mathcal{F} \subset \mathcal{E}$ with $|\mathcal{F}| = 1$. It remains to show that $|A_{\mathcal{F}}| \cdot |X(n)|^{-1} = O(t^2)$ for $|\mathcal{F}| \geq 2$. Since $|A_{\mathcal{F}}|$ could only decrease when $|\mathcal{F}|$ increases, it is sufficient to consider the case $|\mathcal{F}| = 2$. Let $A_{\mathcal{F}} = A_{ij} \cap A_{uv}$ for two different edges $ij, uv \in \mathcal{E}$. If $\{i, j\} \cap \{u, v\} = \emptyset$, it is clear that $|A_{\mathcal{F}}| = |A_{ij}| \cdot |A_{uv}| \cdot |X(n)|^{-1} = t^2 \cdot |X(n)|$. Without restriction, let $j = u$. It holds $(d_1, \ldots, d_N) \in A_{\mathcal{F}}$ if and only if there exist $z_0, z_1 \in \overline{\mathbb{F}}$ with $d_i(z_0) = d_j(z_0) = 0$ and $d_j(z_1) = d_v(z_1) = 0$. If $z_0 = z_1$, one could apply Lemma 5 and gets a cardinality of $|X(n)| \cdot t^2$ for the corresponding subset of $A_{\mathcal{F}}$. Thus, let $z_0 \neq z_1$.

Furthermore, one could assume $z_0, z_1 \in \mathbb{F}$. To see that, let without restriction $2 \leq g_{z_0} \leq \min(n_i, n_j)$. It follows from Lemma 7 (a) that the corresponding subset of A_{ij} has a cardinality of at most $|X(n)| \cdot \sum_{g=2}^{\min(n_i, n_j)} g \cdot \varphi_g \cdot t^{2g} = |X(n)| \cdot \sum_{g=2}^{\min(n_i, n_j)} O(t^g) = |X(n)| \cdot O(t^2)$. Consequently, the cardinality of the corresponding subset of $A_{\mathcal{F}}$ is also at most $|X(n)| \cdot O(t^2)$.

For the case $z_0, z_1 \in \mathbb{F}$, one again applies Lemma 7 (a) and gets a probability of at least $\varphi_1^2 \cdot t^4 = t^2$. In summary, one has $|A_{\mathcal{F}}| \cdot |X(n)|^{-1} = O(t^2)$ for $|\mathcal{F}| \geq 2$ and the proof is finished. □

Now, we come to the situation of pairwise coprimeness. Recall that here all pairs of vertices of $\Gamma(n)$ are connected by an edge, i.e. it holds $E = \frac{N(N-1)}{2}$. The following result is an easy consequence of Theorem 14:

Corollary 4.
For $n := (n_1, \ldots, n_N) \in \mathbb{N}^N$, the set $G(n)$ of N-tuples $(d_1, \ldots, d_N)$ of monic pairwise coprime polynomials $d_i \in \mathbb{F}[z]$ with $\deg(d_i) = n_i$ for $i = 1, \ldots, N$ has the following cardinality:

$$|G(n)| = t^{-(n_1 + \ldots + n_N)} \left(1 - \frac{N(N-1)}{2} \cdot t + O(t^2)\right).$$

Therefore, the probability that $d_1, \ldots, d_N$ are pairwise coprime is equal to

$$1 - \frac{N(N-1)}{2} \cdot t + O(t^2).$$

In the following, we turn to the general graph setup again and proceed as in [20] to firstly achieve an exact formula for $|\Gamma(n)|$ and then, deduce a (better) approximation for it. The following theorem extends a result by [30] from the ring of integers to the ring of polynomials.

With each edge ij of Γ we associate a monic, square-free polynomial $k_{ij}(z) \in \mathbb{F}[z]$. We refer to this as a polynomial labeling of the graph and denote it by $\mathbf{k}$. For each polynomial labeling and vertices $l \in V$, let

$$K_l := \mathrm{lcm}\{k_{ij} \mid ij \in \mathcal{E}_l\}.$$

Then

$$M(n) := \{\mathbf{k} \in \mathbb{F}[z]^E \mid k_{ij} \text{ monic, square-free for } ij \in \mathcal{E}, \ \deg(K_l) \leq n_l, \ l \in V\}$$

is the set of all polynomial labelings $\mathbf{k}$ of Γ satisfying the degree bounds $\deg(K_l) \leq n_l$ for all vertices l. For each monic square-free polynomial p, let $\omega(p)$ denote the number of irreducible factors of p.

Theorem 15.
The cardinality of $\Gamma(n)$ is

$$|\Gamma(n)| = t^{-(n_1 + \ldots + n_N)} \sum_{\mathbf{k} \in M(n)} \prod_{ij \in \mathcal{E}} (-1)^{\omega(k_{ij})} \prod_{l=1}^{N} t^{\deg(K_l)}. \tag{2.7}$$

Proof.
The sets

$$P := \{p \in \mathbb{F}[z] \mid \text{monic, irreducible, } \deg(p) \le \max_{1 \le i \le N} n_i\}$$

and $R := P \times \mathcal{E}$ are finite. For $r = (p, ij) \in R$, define

$$D_r := \{(d_1, ..., d_N) \mid p \mid d_i \text{ and } p \mid d_j\}.$$

Thus, $\Gamma(n) = X(n) \setminus \bigcup_{r \in R} D_r$. From the inclusion-exclusion principle (see Lemma 2), one obtains

$$|\Gamma(n)| = \sum_{S \subset R} (-1)^{|S|} |D_S|, \tag{2.8}$$

where $D_\emptyset = X(n)$ and $D_S := \bigcap_{r \in S} D_r$ for $S \ne \emptyset$.
It remains to determine $|S|$ and $|D_S|$. For each edge ij, define the monic and square-free polynomial

$$k_{ij}^S := \prod_{(p,ij) \in S} p, \tag{2.9}$$

while for each vertex $l \in \mathcal{V}$, we consider the monic and square-free polynomials

$$K_l^S := \text{lcm}\{k_{ij}^S \mid ij \in \mathcal{E}_l\}. \tag{2.10}$$

From the definition of D_S, one obtains $(d_1, \dots, d_N) \in D_S$ if and only if $p \mid \gcd(d_i, d_j)$ is satisfied for all $(p, ij) \in S$. This implies the equivalence:

$$(d_1, ..., d_N) \in D_S \quad \Leftrightarrow \quad k_{ij}^S \mid d_i \text{ and } k_{ij}^S \mid d_j \text{ for } ij \in \mathcal{E}.$$

Note that $k_{ij}^S \mid \gcd\{d_i, d_j\}$ holds for all $ij \in \mathcal{E}$ if and only if $k_{ij}^S \mid d_l$ for all $ij \in \mathcal{E}_l$ and $l \in \mathcal{V}$. Since k_{ij}^S are square-free, this in turn yields the characterization

$$(d_1, ..., d_N) \in D_S \quad \Leftrightarrow \quad K_l^S \mid d_l \text{ for } l \in \mathcal{V}.$$

Thus, one has to count the number of degree n_l monic multiples of a monic polynomial K_l^S, which leads to

$$|D_S| = \prod_{l=1}^N t^{\deg(K_l^S) - n_l} \tag{2.11}$$

if $\deg(K_l^S) \le n_l$ holds for all $l \in \mathcal{V}$ and $|D_S| = 0$ otherwise. To compute $|S|$, note that $\omega(k_{ij}^S)$ coincides with the number of elements $p \in P$ such that $(p, ij) \in S$. Thus

$$|S| = \sum_{ij \in \mathcal{E}} \omega(k_{ij}^S). \tag{2.12}$$

Finally, for each non-empty subset $S \subset R$, equation (2.9) defines a unique polynomial labeling $\mathbf{k}^S \in M(n)$. Conversely, for each $\mathbf{k} \in M(n)$ there exists $S \subset R$ with

$\mathbf{k} = \mathbf{k}^S$. In fact, each polynomial labeling $\mathbf{k} = (k_{ij}|ij \in \mathcal{E}) \in M(n)$ admits a unique factorization into primes

$$k_{ij} = \prod_{p_{ij} \in P_{ij}} p_{ij}$$

for subsets $P_{ij} \subset P$. Defining $S = \bigcup_{ij \in \mathcal{E}} P_{ij} \times \{ij\}$ then yields $k_{ij}^S = k_{ij}$ for all edges $ij \in \mathcal{E}$. Thus, in (2.8) one can sum over polynomial labelings $\mathbf{k}$ instead of summing over S. Moreover, the restriction $\mathbf{k} \in M(n)$ in the sum of (2.7) allows us to use formula (2.11), i.e. we avoid summing up zeros. This completes the proof. $\qquad\square$

In the case that all pairs of vertices of Γ are connected by an edge, one obtains the probability that N monic polynomials are pairwise coprime.

The following remark contains a number theoretical result, which could be concluded from the preceding theorem. It is not relevant for our further considerations but seems to be interesting on its own.

Remark 9.
For $N = 2$ and $E = 1$, formula (2.7) has the following form:

$$t^{-n_1-n_2} \sum_{g=0}^{\min(n_1,n_2)} \left(\sum_{\substack{k \text{ monic, square-free} \\ \deg(k)=g}} (-1)^{\omega(k)} \right) t^{2g} =$$

$$= t^{-n_1-n_2} \left(1 - t + \sum_{g=2}^{\min(n_1,n_2)} \left(\sum_{\substack{k \text{ monic, square-free} \\ \deg(k)=g}} (-1)^{\omega(k)} \right) t^{2g} \right).$$

Recall that the number of coprime pairs of monic polynomials is $t^{-n_1-n_2}(1-t)$; see Lemma 5. Thus, one obtains the combinatorical identity:

$$\sum_{g=2}^{\min(n_1,n_2)} \sum_{\substack{k \text{ monic, square-free} \\ \deg(k)=g}} (-1)^{\omega(k)} t^{2g} = \sum_{g=2}^{\min(n_1,n_2)} (|E(g)| - |U(g)|) t^{2g} = 0,$$

where

$$|E(g)| := \{(p_1, \ldots, p_{2r}) \mid r \in \mathbb{N}, p_i \neq p_j \text{ monic, irreducible, } \sum_{i=1}^{2r} \deg(p_i) = g\}$$

$$|U(g)| := \{(p_1, \ldots, p_{2r-1}) \mid r \in \mathbb{N}, p_i \neq p_j \text{ monic, irreducible, } \sum_{i=1}^{2r-1} \deg(p_i) = g\}$$

are the numbers of monic, square-free polynomials with an even or odd number of *irreducible factors, respectively. For $n \geq 2$, it follows:*

$$(|E(n)| - |U(n)|)t^{2n} = \sum_{g=2}^{n}(|E(g)| - |U(g)|)t^{2g} - \sum_{g=2}^{n-1}(|E(g)| - |U(g)|)t^{2g} =$$

$$= 0 - 0 = 0,$$

i.e. $|E(n)| = |U(n)|$ *for every* $n \geq 2$.

In words, for $n \geq 2$, there are as many monic polynomials of degree n with an even number of distinct prime factors as with an odd number of distinct prime factors. This result was firstly proven by Carlitz [6] in 1932. This means that it is quite old but we provided a new proof for it along the way.

Returning to our actual goal, there is the problem that for $N > 2$, the formula of Theorem 15 is very difficult to evaluate. However, if the degree of one of the polynomials is at least as large as the sum of the other degrees, the computation could be reduced to a computation with lower degrees. This fact is explicitly stated in the following corollary:

Corollary 5.
Let $n_1, ..., n_N, h \in \mathbb{N}$. *Then:*

$$|\Gamma(n_1, ..., n_N)| = t^{-h}|\Gamma(n_1 - h, n_2, ..., n_N)| \quad \text{if} \quad n_1 = h + n_2 + ... + n_N.$$

Proof.
For $\mathbf{k} \in M(n)$ it holds:

$$\deg(K_1) \leq \sum_{ij \in \mathcal{E}_1} \deg(k_{ij}) \leq \sum_{l=2}^{N} \deg(k_{1l}) \leq \sum_{l=2}^{N} \deg(K_l) \leq \sum_{l=2}^{N} n_l \leq n_1. \quad (2.13)$$

The first and the third inequality follow because K_l is the least common multiple of the corresponding k_{ij}. The fourth inequality holds since $\mathbf{k} \in M(n)$. Finally, the last inequality holds because of the assumption $n_1 = h + n_2 + ... + n_N$.
From (2.13) it follows that in the given situation increasing n_1 does not increase the number of elements in $M(n)$ because $\deg(K_l)$ are stronger restricted by $n_2, ..., n_N$ than by n_1. Thus, the only expression in (2.7) that changes when increasing n_1 is t^{-n_1}, which causes the factor t^{-h}. □

Next, Theorem 15 is used to give an alternative proof for the asymptotic formula from Theorem 14.

Proof.
To prove this result, first sort the elements of $M(n)$ with respect to the degrees of the entries of the vector $\mathbf{k} = (k_1, ..., k_E)$.
To this end, for each vector of non-negative integers $\mathbf{g} := (g_1, ..., g_E)$, define $M(n, \mathbf{g}) := \{\mathbf{k} \in M(n) \mid \deg(k_m) = g_m \text{ for } 1 \leq m \leq E\}$. Let A be the set of all $\mathbf{g}$

with $M(n, \mathbf{g}) \neq \emptyset$. Note that the degree bounds for $M(n)$ ensure that A is finite. One achieves:

$$|\Gamma(n)| = t^{-(n_1+\ldots+n_N)} \sum_{\mathbf{g} \in A} \sum_{\mathbf{k} \in M(n,\mathbf{g})} \prod_{ij \in \mathcal{E}} (-1)^{\omega(k_{ij})} \prod_{l=1}^{N} t^{\deg(K_l)}.$$

Starting with small values for the entries of $\mathbf{g}$ the first summands are computed. For $\mathbf{g} = (0, \ldots, 0)$, i.e. $\mathbf{k} = (1, \ldots, 1)$, one gets the summand 1 because of $\omega(1) = 0$ and $K_l = 1$ for $l = 1, \ldots, N$. If $g_{m_0} = 1$ for exactly one $1 \leq m_0 \leq E$ and $g_m = 0$ for $m \neq m_0$, there are $|\mathbb{F}| = 1/t$ possibilities for the linear polynomial k_{m_0} and E possibilities for the choice of m_0. Moreover, $\omega(k_{m_0}) = 1$, so that these summands have negative sign. As k_{m_0} is relevant for exactly those K_l for which its associated edge is terminating at l, there are exactly two K_l which are of degree 1. Hence, the resulting sum of these terms is equal to $-E \cdot \frac{1}{t} \cdot t^2 = -E \cdot t$.

Thus, one only has to show that each of the remaining summands behaves asymptotically as $O(t^2)$, which is done by showing

$$R(\mathbf{g}) := \sum_{\mathbf{k} \in M(n,\mathbf{g})} \prod_{l=1}^{N} t^{\deg(K_l)} = O(t^2)$$

for every fixed $\mathbf{g}$ for which the sum of the entries of $\mathbf{g}$ is at least two.

This will be done by induction with respect to E.

For $E = 1$, note that $\mathbf{g}$ and $\mathbf{k} = k_{12}$ are scalar. Moreover, $K_1 = K_2 = k_{12}$. Therefore, $R(\mathbf{g}) = 0$ if $\mathbf{g} > \min(n_1, n_2)$ and otherwise,

$$R(\mathbf{g}) \leq \sum_{\mathbf{k} \text{ monic, } \deg(\mathbf{k})=\mathbf{g}} t^{2 \deg(\mathbf{k})} = \left(\frac{1}{t}\right)^{\mathbf{g}} \cdot t^{2\mathbf{g}} = t^{\mathbf{g}} = O(t^2) \text{ for } \mathbf{g} \geq 2.$$

This computation starts with an inequality since the condition that $\mathbf{k}$ has to be square-free is dropped. The first equality follows from the fact that there are $(1/t)^{\mathbf{g}}$ monic polynomials of degree $\mathbf{g}$.

Next, we take the step from $E - 1$ to E.

To this end, choose one of the smallest entries of $\mathbf{g}$ and denote it without loss of generality by g_E. Then, the edge with which k_E is associated – in the following denoted by ij – is taken away form the original graph and thus, a graph with $E - 1$ edges is achieved. In the following, the index $(E - 1)$ above an expression means that it belongs to a graph with $E - 1$ edges; in the same way, we use the index (E). Similarly, $\mathbf{k}^{(E-1)}$ and $\mathbf{g}^{(E-1)}$ should denote the vectors consisting of the first $E - 1$ entries of $\mathbf{k}$ and $\mathbf{g}$, respectively.

The degrees of the K_l can never increase, when taking an edge away. Therefore, $\mathbf{k} \in M(n, \mathbf{g})$ implies $\mathbf{k}^{(E-1)} \in M^{(E-1)}(n, \mathbf{g}^{(E-1)})$. Next, we set

$$W_i := \gcd(K_i^{(E-1)}, k_E) \quad \text{and} \quad W_j := \gcd(K_j^{(E-1)}, k_E).$$

Moreover, let

$$B_{v_i,v_j}^{(E-1)} := \{\mathbf{k}^{(E-1)} \in M^{(E-1)}(n, \mathbf{g}^{(E-1)}) \mid \deg(K_i^{(E-1)}) = v_i, \ \deg(K_j^{(E-1)}) = v_j\},$$

$$B_{v_i,v_j,w_i,w_j}^{(E)} := \{\mathbf{k} \in M^{(E)}(n, \mathbf{g}) \mid \mathbf{k}^{(E-1)} \in B_{v_i,v_j}^{(E-1)}, \deg(W_i) = w_i, \deg(W_j) = w_j\}.$$

It follows

$$R(\mathbf{g}) \le \sum_{v_i,v_j,w_i,w_j \le \max(n_i,n_j)} \ \sum_{\mathbf{k} \in B_{v_i,v_j,w_i,w_j}^{(E)}} \ \prod_{l=1}^{N} t^{\deg(K_l^{(E)})}.$$

The number of summands in the first sum is finite and thus one only has to show that for any fixed v_i, v_j, w_i, w_j the following is true:

$$\sum_{\mathbf{k} \in B_{v_i,v_j,w_i,w_j}^{(E)}} \ \prod_{l=1}^{N} t^{\deg(K_l^{(E)})} = O(t^2).$$

To do this one computes

$$K_i^{(E)} = \operatorname{lcm}(K_i^{(E-1)}, k_E) = \frac{K_i^{(E-1)} \cdot k_E}{W_i}.$$

Consequently, one has $\deg(K_i^{(E)}) = \deg(K_i^{(E-1)}) + g_E - w_i$ and $\deg(K_j^{(E)}) = \deg(K_j^{(E-1)}) + g_E - w_j$, analogously. For $l \notin \{i,j\}$, it holds $K_l^{(E)} = K_l^{(E-1)}$ because nothing changes at the associated vertices. It follows:

$$\sum_{\mathbf{k} \in B_{v_i,v_j,w_i,w_j}^{(E)}} \ \prod_{l=1}^{N} t^{\deg(K_l^{(E)})} = \sum_{\mathbf{k} \in B_{v_i,v_j,w_i,w_j}^{(E)}} \ \prod_{l=1}^{N} t^{\deg(K_l^{(E-1)})} \cdot t^{2g_E - w_i - w_j}. \qquad (2.14)$$

Here, the product $\prod_{l=1}^{N} t^{\deg(K_l^{(E-1)})}$ is independent of k_E and $t^{2g_E - w_i - w_j}$ is independent of $\mathbf{k} \in B_{v_i,v_j,w_i,w_j}^{(E)}$.

Next, for $\mathbf{k}^{(E-1)} \in B_{v_i,v_j}^{(E-1)}$, an upper bound for the number of polynomials k_E such that $\mathbf{k} \in B_{v_i,v_j,w_i,w_j}^{(E)}$ should be determined. $\mathbf{k}^{(E-1)}$ uniquely determines $K_i^{(E-1)}$ and since W_i is a divisor of $K_i^{(E-1)}$ of degree w_i, there are only finitely many possibilities for W_i. Define C as this number of possibilities for W_i. One knows that k_E has to be a multiple of W_i of degree g_E. Thus, for every W_i, there are at most $t^{w_i - g_E}$ possibilities for k_E.

Using this and the fact that the product in (2.14) is independent of k_E, it follows for the expression in (2.14):

$$\sum_{k \in B^{(E)}_{v_i,v_j,w_i,w_j}} \prod_{l=1}^{N} t^{\deg(K_l^{(E)})} \leq t^{2g_E - w_i - w_j} \cdot C \cdot t^{w_i - g_E} \sum_{k^{(E-1)} \in B^{(E-1)}_{v_i,v_j}} \prod_{l=1}^{N} t^{\deg(K_l^{(E-1)})}$$

$$= Ct^{g_E - w_j} \sum_{k^{(E-1)} \in B^{(E-1)}_{v_i,v_j}} \prod_{l=1}^{N} t^{\deg(K_l^{(E-1)})}$$

$$\leq C \cdot R(g^{(E-1)})$$

because $w_j \leq g_E$ since $W_j \mid k_E$. Now, we distinguish three cases:

Case 1: The sum of the entries of $g^{(E-1)}$ is at least two.
Then, $R(g^{(E-1)})$ is $O(t^2)$ per induction and we are done.

Case 2: $g^{(E-1)}$ has a component that is equal to zero.
Here, g_E must be zero since it was choosen to be one of the smallest entries. But then the sum of the entries of $g^{(E-1)}$ is equal to the sum of the entries of $g^{(E)}$ and thus, in particular, at least two. Consequently, we are done, too.

Case 3: $E = 2$ and $g^{(E-1)} = g_1 = 1$.
Then $g_E = g_2 \leq 1$. If $g_2 = 0$, we argue as before. If $g_2 = 1$ and the two edges of the graph meet at a vertex, one gets

$$R(g) = R(1,1) \leq \sum_{\substack{k_1, k_2 \text{ monic} \\ \deg(k_m)=1}} t^{2+\deg(\mathrm{lcm}(k_1,k_2))} =$$

$$= \sum_{\substack{k_1 = k_2 \text{ monic} \\ \deg(k_m)=1}} t^3 + \sum_{\substack{k_1 \neq k_2 \text{ monic} \\ \deg(k_m)=1}} t^4 =$$

$$= \frac{1}{t} \cdot t^3 + \frac{1}{t} \cdot \left(\frac{1}{t} - 1\right) \cdot t^4 = O(t^2).$$

If $g_2 = 1$ and the two edges of the graph are isolated, one gets

$$R(g) = R(1,1) \leq \sum_{\substack{k_1, k_2 \text{ monic} \\ \deg(k_m)=1}} t^4 = t^2$$

since there are two K_l that coincide with k_1 and k_2, respectively. Moreover, there are t^{-2} pairs of monic polynomials of degree one.
Thus, this case is done as well and our proof is complete. $\qquad\qquad\square$

Although this proof is far more involved than that following Theorem 14, it has the advantage that in principle, it is possible to compute the coefficients of t^j for $j \geq 2$

with the same method. But with increasing j the computational effort becomes very large because too many different cases of graph structures have to be distinguished. Moreover, the possible subgraphs that have to be considered depend on the original graph Γ. In the preceding proof, the only subgraph that needed exact computation was that with only one edge, which is contained in every nonempty graph. However, for the computation of higher coefficients, one has to look at subgraphs with more than one edge and so not every graph with E edges contains the same set of these subgraphs, which leads to a very complicated case distinction. Therefore, in the following, we only focuse on the complete graph, where N vertices are connected pairwisely, i.e. the case of pairwise coprimeness, and compute the coefficient of t^2 for this case.

Theorem 16.
Let $n_1, ..., n_N \in \mathbb{N}$ and $N_1 := |l \in \{1, ..., N\} \mid n_l = 1|$. Then, the probability that N monic polynomials over $\mathbb{F}$ of degrees $n_1, ..., n_N$ are pairwise coprime is equal to

$$1 - \frac{N(N-1)}{2} \cdot t + \frac{1}{24}(N-1)(N-2)(3N^2 + 11N - 12N_1) \cdot t^2 + O(t^3).$$

Proof.
Let G be a graph with N vertices, which are pairwisely connected by an edge and let $|G(n)|$ be the number of N-tuples of monic pairwisely coprime polynomials over $\mathbb{F}$ of degrees $n_1, \ldots, n_N$. Moreover, Γ should be any subgraph of G, whose number of edges is equal to E. To prove the result, we first consider the general graph Γ. As in the proof of the preceding theorem and with the same notation as there, one gets

$$|\Gamma(n)| = t^{-(n_1 + \ldots + n_N)} \sum_{\mathbf{g} \in A} \sum_{\mathbf{k} \in M(n, \mathbf{g})} \prod_{ij \in \mathcal{E}} (-1)^{\omega(k_{ij})} \prod_{l=1}^{N} t^{\deg(K_l)}.$$

Now consider G, i.e. the case $E = \frac{N(N-1)}{2}$. Starting with small values for the entries of $\mathbf{g}$, the first summands are computed.
As before, the sum of the terms with $\mathbf{g} = (0, \ldots, 0)$ or $g_{m_0} = 1$ for exactly one $1 \le m_0 \le E$ and $g_m = 0$ for $m \ne m_0$ is equal to $1 - E \cdot \frac{1}{t} \cdot t^2 = 1 - E \cdot t$.
Note that for all summands computed so far, every $\mathbf{k}$ lies in $M(n, \mathbf{g})$ since $\deg(K_l) \le 1$ in all considered cases. Next, look at the summands whose sum of the entries of $\mathbf{g}$ is equal to 2. All summands for which $g_{m_0} = 2$ for exactly one $1 \le m_0 \le E$ and $g_m = 0$ for $m \ne m_0$ have modulus t^4. Since they have negative sign if $k_{m_0} \in U(2)$ and positive sign if $k_{m_0} \in E(2)$, it follows from $|E(2)| = |U(2)|$ that these summands add up to zero. Hence, in this case, it does not matter whether $\mathbf{k}$ lies in $M(n, \mathbf{g})$ or not since this depends only on m_0 and not on k_{m_0} itself.
Now consider the summands for which two entries of $\mathbf{g}$ are equal to one, and the other entries are equal to zero. If the corresponding edges of the nonzero entries

have a vertex l in common, the summand has the value

$$\sum_{\substack{k_1, k_2 \text{ monic} \\ \deg(k_m)=1}} t^{2+\deg(\text{lcm}(k_1,k_2))} = \sum_{\substack{k_1 = k_2 \text{ monic} \\ \deg(k_m)=1}} t^3 + \sum_{\substack{k_1 \neq k_2 \text{ monic} \\ \deg(k_m)=1}} t^4 = \tag{2.15}$$

$$= \frac{1}{t} \cdot t^3 + \frac{1}{t} \cdot \left(\frac{1}{t} - 1\right) \cdot t^4 = 2t^2 - t^3 \tag{2.16}$$

if $n_l \geq 2$ and t^2 if $n_l = 1$ since the summands of the second sum lie not in $M(n, \mathbf{g})$ if $\deg(K_l) = 2 > n_l$. For such an "angle", there are $N \cdot \binom{N-1}{2}$ possibilities, N for the apex and $\binom{N-1}{2}$ for the two sides of the angle.
If those two edges are isolated, the summand has the value

$$\sum_{\substack{k_1, k_2 \text{ monic} \\ \deg(k_m)=1}} t^4 = t^2.$$

For this case, there are $\binom{N}{4}$ possibilities to choose the four involved vertices and 3 possibilities to connect two of them, pairwisely.
In summary, all summands whose sum of the entries of $\mathbf{g}$ is equal to two contribute the value

$$\left(\binom{N-1}{2}(2(N-N_1)+N_1)+3\cdot\binom{N}{4}\right)\cdot t^2 + O(t^3) =$$

$$\left(\frac{(N-1)(N-2)}{2}(2N-N_1)+\frac{N(N-1)(N-2)(N-3)}{8}\right)\cdot t^2 + O(t^3) =$$

$$\frac{(N-1)(N-2)}{8}(N^2+5N-4N_1)\cdot t^2 + O(t^3). \tag{2.17}$$

If $\mathbf{g}$ contains three ones where the corresponding edges form a triangle and zeros in the other entries, one gets

$$-\frac{1}{t}\cdot t^3 - \frac{3}{t}\cdot\left(\frac{1}{t}-1\right)\cdot t^5 - \frac{1}{t}\cdot\left(\frac{1}{t}-1\right)\cdot\left(\frac{1}{t}-2\right)\cdot t^6 = -t^2 + O(t^3).$$

Here, the first summand of the left side of the equation covers the case that three, the second summand that two and the third summand that none of the three entries of $\mathbf{k}$ that contain a linear polynomial are identical. Moreover, there are $\binom{N}{3}$ possibilities for such a triangle.
Adding these summands to (19), one gets

$$\frac{(N-1)(N-2)}{8}(N^2+5N-4N_1-4N/3)\cdot t^2 + O(t^3) =$$

$$\frac{(N-1)(N-2)}{24}(3N^2+11N-12N_1)\cdot t^2 + O(t^3).$$

Now we turn to the general graph Γ again and show that

$$R(\mathbf{g}) := \sum_{\mathbf{k} \in M(n,\mathbf{g})} \prod_{l=1}^{N} t^{\deg(K_l)} = O(t^3)$$

for every fixed $\mathbf{g}$ for which the sum of the entries of $\mathbf{g}$ is at least three and Γ is no triangle.

Analogously to the proof of the preceding theorem, one gets $R(\mathbf{g}) \leq t^{\mathbf{g}}$ for $E = 1$ and $R(\mathbf{g}^{(E)}) \leq C \cdot R(\mathbf{g}^{(E-1)})$ with $C \in \mathbb{N}$ for $E \geq 2$.

However, the number of cases that have to be distingushed is larger:

Case 1: The sum of the entries of $\mathbf{g}^{(E-1)}$ is at least three and no triangle.
Then, $R(\mathbf{g}^{(E-1)})$ is $O(t^3)$ per induction and we are done.

Case 2: $\mathbf{g}^{(E-1)} = (1,1,1)$ and $\Gamma^{(E-1)}$ is a triangle.
This case can be avoided: It holds $\mathbf{g}^{(E)} = (1,1,1,1)$ since $\mathbf{g}^{(E)} = (1,1,1,0)$ would mean that $\Gamma^{(E)}$ is a triangle, too, because an edge ij with labelling $k_{ij} = 1$ could be treated like it would not exist. Therefore, one of the vertices of the triangle has an third edge which connects it with the additional vertex. Since all entries of $\mathbf{g}^{(E)}$ are identical, one can take away an arbitrary edge in our process of induction. If one takes away one of the edges which form the triangle, the resulting $\Gamma^{(E-1)}$ is not a triangle any more.

It remains to consider all possible cases for which the sum of the entries of $\mathbf{g}^{(E-1)}$ is smaller than three but the sum of the entries of $\mathbf{g}^{(E)}$ is at least three and $\Gamma^{(E)}$ is no triangle. First, one excludes zero entries in these vectors (case 3) and then, considers $\mathbf{g}^{(E-1)} = (1,1)$ (case 4) and $\mathbf{g}^{(E-1)} = 2$ (cases 5 and 6).

Case 3: $\mathbf{g}^{(E-1)}$ has a component that is equal to zero.
Note that this case corresponds to case 2 of the preceding proof.
Here, g_E must be zero since it was choosen to be one of the smallest entries. Thus, $\Gamma^{(E-1)}$ and $\Gamma^{(E)}$ could be treated as being identical and hence, $\Gamma^{(E-1)}$ fulfills the conditions of case 1. Consequently, we are done, too.

Case 4: $\mathbf{g}^{(E)} = (1,1,1)$ and $\Gamma^{(E)}$ is no triangle.
Case 4a: $\Gamma^{(E)}$ consists of three isolated edges:

$$R(\mathbf{g}) \leq \left(\frac{1}{t} \right)^3 \cdot t^6 = t^3 = O(t^3).$$

Case 4b: $\Gamma^{(E)}$ consists of an isolated edge and an angle (see (2.15)):

$$R(\mathbf{g}) \leq \frac{1}{t} \cdot t^2 \cdot (2t^2 - t^3) = O(t^3).$$

Case 4c: $\Gamma^{(E)}$ consists of three edges forming one line:

$$R(g) \leq \frac{1}{t} \cdot t^4 + \frac{2}{t} \left(\frac{1}{t} - 1\right) \cdot t^5 + \frac{1}{t} \left(\frac{1}{t} - 1\right)^2 \cdot t^6 = O(t^3).$$

The first summand covers the case that all linear polynomials in k are identical, the second summand the case that the polynomial of the edge in the middle coincides with one of the others and the third polynomial is different and the third summand the case that the polynomial in the middle is different from the other two polynomials.

Case 4d: $\Gamma^{(E)}$ consists of three edges that meet at one vertex:

$$R(g) \leq \frac{1}{t} \cdot t^4 + \frac{3}{t} \left(\frac{1}{t} - 1\right) \cdot t^5 + \frac{1}{t} \left(\frac{1}{t} - 1\right) \left(\frac{1}{t} - 2\right) \cdot t^6 = O(t^3).$$

The first summand covers the case that all linear polynomials in k are identical, the second summand the case that exactly two of them are identical and the third summand the case that all polynomials are different.

Case 5: $g^{(E)} = (2,1)$.
Since we are considering upper bounds for $R(g)$ in the following, we can drop the condition that the quadratic polynomials have to be square-free.
Case 5a: $\Gamma^{(E)}$ consists of two isolated edges:

$$R(g) \leq \left(\frac{1}{t}\right)^3 \cdot t^6 = O(t^3).$$

Case 5b: $\Gamma^{(E)}$ consists of an angle:

$$R(g) \leq \frac{1}{t} \left(\frac{1}{t} - 1\right) \cdot t^5 + \frac{1}{t^3} \cdot t^6 = O(t^3).$$

The first summand covers the case that the linear polynomial divides the quadratic polynomial, the second summand the other case.

Case 6: $g^{(E)} = (2,2)$.
Case 6a: G consists of two isolated edges:

$$R(g) \leq \left(\frac{1}{t}\right)^4 \cdot t^8 = O(t^3).$$

Case 6b: G consists of an angle:

$$R(g) \leq \frac{1}{t^2} \cdot t^6 + \frac{1}{t^4} \cdot t^7 = O(t^3).$$

The first summand covers the case that the two quadratic polynomials are identical, the second summand the other case.

It follows that $R(\mathbf{g}) = O(t^3)$ for every fixed $\mathbf{g}$ for which the sum of the entries of $\mathbf{g}$ is at least three and Γ is no triangle. Consequently,

$$|G(n)| = t^{n_1 + \ldots + n_N}.$$

$$\left(1 - \frac{N(N-1)}{2} \cdot t + \frac{1}{24}(N-1)(N-2)(3N^2 + 11N - 12N_1) \cdot t^2 + O(t^3)\right).$$

$\square$

So far, we used the uniform probability distribution and fixed the degrees of the considered polynomials. In the following, this result should be compared with the natural density of pairwise coprime polynomials with arbitrary degrees. To this end, we follow the lines of the proof of Theorem 1 from [17]. The following Theorem was also proven in [18, Corollary 1].

Theorem 17.
The natural density of pairwise coprimeness for N arbitrary polynomials $d_1, \ldots, d_N \in \mathbb{F}[z]$ is equal to

$$\prod_{j=1}^{\infty} \left((1 - t^j)^{N-1}(1 + (N-1)t^j)\right)^{\varphi_j}.$$

Proof.
From Theorem 6, one knows that $d_1, \ldots, d_N$ are pairwise coprime if and only if the matrix

$$\mathcal{D}_N := \begin{bmatrix} d_1 & d_2 & 0 & \cdots & 0 \\ 0 & d_2 & d_3 & \ddots & \vdots \\ \vdots & \ddots & \ddots & \ddots & 0 \\ 0 & \cdots & 0 & d_{N-1} & d_N \end{bmatrix}$$

is left prime. According to Remark 3 (b), this holds if and only if the size $N - 1$ minors of $\mathcal{D}_N$ are coprime.

In the following, the notation of Definition 10 is used. Let M_n be the set of all tuples $(d_1, \ldots, d_N) \in \mathbb{F}[z]^N$ for which $d_i \in \{f_0, \ldots, f_n\}$ for $i = 1, \ldots, N$. Furthermore, let $\hat{P}$ be the set of all (monic) irreducible polynomials in $\mathbb{F}[z]$ and P a finite subset of $\hat{P}$. Moreover, E_P should denote the set of all tuples $(d_1, \ldots, d_N) \in \mathbb{F}[z]^N$ for which the gcd of all size $N - 1$ minors of $\mathcal{D}_N$ is coprime with all elements in P. Consequently, we are interested in the probability that $(d_1, \ldots, d_N) \in \mathbb{F}[z]^N$ lies in $E := \bigcap_P E_P$; i.e., to obtain the natural density one has to determine $\lim_{n\to\infty} \frac{|E \cap M_n|}{|M_n|}$.

In a first step, one computes the probability that $(d_1, \ldots, d_N) \in M_n$ lies in E_P. To this end, one defines $f_P := \prod_{f \in P} f$ and $d_P := \deg(f_P)$. Next, consider the projection

$$M_n \to M_n/(f_P) = \prod_{f \in P} M_n/(f) :$$

$$(d_1, \ldots, d_N) \mapsto (d_1, \ldots, d_N)/(f_P) = \prod_{f \in P} (d_1, \ldots, d_N)/(f),$$

which applies the canonical projection modulo f_P ($\mathbb{F}[z] \to \mathbb{F}[z]/(f_P)$) to each entry of $(d_1, \ldots, d_N)$. For $(d_1, \ldots, d_N) \in M_n$ holds:

$$(d_1, \ldots, d_N) \in E_P \Leftrightarrow \forall f \in P \; \exists \text{ fullsize minor of } \mathcal{D}_N \text{ that is not divided by } f$$

$$\Leftrightarrow \forall f \in P \; \exists \text{ fullsize minor of } \mathcal{D}_N \text{ that is nonzero in } (\mathbb{F}[z]/(f))^{N-1 \times N}$$

$$\Leftrightarrow \forall f \in P : \; \mathcal{D}_N/(f) \text{ has full rank in } (\mathbb{F}[z]/(f))^{N-1 \times N} \simeq (\mathbb{F}^{\deg f})^{N-1 \times N},$$

where $\mathbb{F}^{\deg f}$ denotes the field with $t^{-\deg(f)}$ elements. Note that the matrix

$$\mathcal{D}_N/(f) = \begin{bmatrix} a_1 & a_2 & 0 & \cdots & 0 \\ 0 & a_2 & a_3 & \ddots & \vdots \\ \vdots & \ddots & \ddots & \ddots & 0 \\ 0 & \cdots & 0 & a_{N-1} & a_N \end{bmatrix} \in (\mathbb{F}^{\deg f})^{N-1 \times N}$$

has full rank if and only if $a_i = 0$ for at most one $i \in \{1, \ldots, N\}$. The probability for that is equal to $(1 - t^{\deg(f)})^N + N t^{\deg(f)}(1 - t^{\deg(f)})^{N-1}$.

First, suppose that t^{-d_P} divides $|\{f_0, \ldots, f_n\}| = n + 1$, i.e. $n = mt^{-d_P} - 1$ for some $m \in \mathbb{N}$. Then, one could write $\{f_0, \ldots, f_n\} = \{f_s(z)z^{d_P} + f_r(z) \mid 0 \leq s \leq m-1, \, 0 \leq r \leq t^{-d_P} - 1\}$. One has $\{f_r \mid 0 \leq r \leq t^{-d_P} - 1\} \simeq \mathbb{F}[z]/(f_P)$ and $f_s(z)z^{d_P} + f_r(z)$ mod $f_P(z) = f_s(z)z^{d_P}$ mod $f_P(z) + f_r(z) = \hat{f}_s(z) + f_r(z)$ where $\hat{f}_s(z) := f_s(z)z^{d_P}$ mod $f_P(z) \in \mathbb{F}[z]/(f_P)$. Hence, for every fixed s the canonical projection is bijective and on $\{f_0, \ldots, f_n\}$ it is m-to-one. In summary, one obtains

$$|E_P \cap M_n| = m^N \cdot \prod_{f \in P} t^{-N \deg(f)}((1 - t^{\deg(f)})^N + N t^{\deg(f)}(1 - t^{\deg(f)})^{N-1}) =$$

$$= (mt^{-d_P})^N \cdot \prod_{f \in P}(1 - t^{\deg(f)})^N + N t^{\deg(f)}(1 - t^{\deg(f)})^{N-1}.$$

Since $mt^{-d_P} = n + 1$, i.e. $(mt^{-d_P})^N = |M_n|$, it follows

$$\frac{|E_P \cap M_n|}{|M_n|} = \prod_{f \in P}(1 - t^{\deg(f)})^N + N t^{\deg(f)}(1 - t^{\deg(f)})^{N-1}.$$

Now, suppose $n \in \mathbb{N}$ arbitrary. By division with remainder, we get $n + 1 = mt^{-d_P} + r$ with $0 \leq r < t^{-d_P}$. One defindes $\hat{n} := n - r = mt^{-d_P} - 1$. Since

$$\lim_{n \to \infty} \frac{|E_P \cap (M_n \setminus M_{\hat{n}})|}{|M_n|} \leq \lim_{n \to \infty} \frac{|M_n| - |M_{\hat{n}}|}{|M_n|} =$$

$$= \lim_{n \to \infty} \frac{(n+1)^N - (n+1-r)^N}{(n+1)^N} = 0,$$

one has

$$\lim_{n\to\infty}\frac{|E_P\cap M_n|}{|M_n|}=\lim_{n\to\infty}\frac{|E_P\cap M_{\hat{n}}|+|E_P\cap(M_n\setminus M_{\hat{n}})|}{|M_n|}=\lim_{n\to\infty}\frac{|E_P\cap M_{\hat{n}}|}{|M_n|}$$

$$=\lim_{n\to\infty}\frac{(n-r+1)^N\prod_{f\in P}(1-t^{\deg(f)})^N+Nt^{\deg(f)}(1-t^{\deg(f)})^{N-1}}{(n+1)^N}=$$

$$=\prod_{f\in P}(1-t^{\deg(f)})^N+Nt^{\deg(f)}(1-t^{\deg(f)})^{N-1}.$$

Easy computation leads to

$$(1-t^{\deg(f)})^N+Nt^{\deg(f)}(1-t^{\deg(f)})^{N-1}=(1-t^{\deg(f)})^{N-1}(1+(N-1)t^{\deg(f)})$$

$$=\left(1-(N-1)t^{\deg(f)}+\binom{N-1}{2}t^{2\deg(f)}+\sum_{k=3}^{N}\tilde{\alpha}_k t^{k\cdot\deg(f)}\right)(1+(N-1)t^{\deg(f)})$$

$$=1+\frac{(N-1)(N-2)-2(N-1)^2}{2}t^{2\deg(f)}+\sum_{k=3}^{N}\alpha_k t^{k\cdot\deg(f)}$$

$$=1-\binom{N}{2}t^{2\deg(f)}+\sum_{k=3}^{N}\alpha_k t^{k\cdot\deg(f)},$$

with $\tilde{\alpha}_k,\alpha_k\in\mathbb{N}$, which are independent of $\deg(f)$. Define $H_f=\mathbb{F}[z]^N\setminus E_f$. Then

$$\lim_{n\to\infty}\frac{|H_f\cap M_n|}{|M_n|}=1-\lim_{n\to\infty}\frac{|E_f\cap M_n|}{|M_n|}=\binom{N}{2}t^{2\deg(f)}+\sum_{k=3}^{N}\alpha_k t^{k\cdot\deg(f)}.$$

Set $\alpha:=\max(\alpha_3,\dots,\alpha_N)$ and let P_g be the set of all irreducible polynomials with degree at most g. Then $E_{P_g}\setminus E\subset\bigcup_{f\in\hat{P}\setminus P_g}H_f$ and consequently,

$$\limsup_{n\to\infty}\frac{|(E_{P_g}\setminus E)\cap M_n|}{|M_n|}\le\limsup_{n\to\infty}\frac{|(\bigcup_{f\in\hat{P}\setminus P_g}H_f)\cap M_n|}{|M_n|}$$

$$\le\limsup_{n\to\infty}\frac{\sum_{f\in\hat{P}\setminus P_g}|H_f\cap M_n|}{|M_n|}$$

$$\le\sum_{f\in\hat{P}\setminus P_g}\limsup_{n\to\infty}\frac{|H_f\cap M_n|}{|M_n|}=\sum_{f\in\hat{P}\setminus P_g}\left(\binom{N}{2}t^{2\deg(f)}+\sum_{k=3}^{N}\alpha_k t^{k\cdot\deg(f)}\right)=$$

$$=\sum_{j=g+1}^{\infty}\varphi_j\left(\binom{N}{2}t^{2j}+\sum_{k=3}^{N}\alpha_k t^{k\cdot j}\right)\le\sum_{j=g+1}^{\infty}t^{-j}\left(\binom{N}{2}t^{2j}+\sum_{k=3}^{N}\alpha_k t^{k\cdot j}\right)$$

$$\le\sum_{j=g+1}^{\infty}\binom{N}{2}t^{j}+\sum_{k=3}^{N}\alpha_k t^{(k-1)\cdot j}\le\binom{N}{2}\frac{t^{g+1}}{1-t}+\alpha(N-2)\sum_{j=g+1}^{\infty}t^{2j}\overset{g\to\infty}{\longrightarrow}0.$$

Since $E \cap M_n = E_{P_g} \cap M_n \setminus ((E_{P_g} \setminus E) \cap M_n)$, one obtains

$$\liminf_{n \to \infty} \frac{|E \cap M_n|}{|M_n|} \geq \liminf_{n \to \infty} \frac{|E_{P_g} \cap M_n|}{|M_n|} - \limsup_{n \to \infty} \frac{|(E_{P_g} \setminus E) \cap M_n|}{|M_n|}$$

$$\geq \lim_{n \to \infty} \frac{|E_{P_g} \cap M_n|}{|M_n|} - \binom{N}{2} \frac{t^{g+1}}{1-t} + \alpha(N-2) \frac{t^{2(g+1)}}{1-t^2}$$

as well as

$$\limsup_{n \to \infty} \frac{|E \cap M_n|}{|M_n|} \leq \limsup_{n \to \infty} \frac{|E_{P_g} \cap M_n|}{|M_n|} - \liminf_{n \to \infty} \frac{|(E_{P_g} \setminus E) \cap M_n|}{|M_n|}$$

$$\leq \lim_{n \to \infty} \frac{|E_{P_g} \cap M_n|}{|M_n|}.$$

It follows

$$\lim_{n \to \infty} \frac{|E \cap M_n|}{|M_n|} = \lim_{g \to \infty} \lim_{n \to \infty} \frac{|E_{P_g} \cap M_n|}{|M_n|} =$$

$$= \lim_{g \to \infty} \prod_{f \in P_g} \left((1 - t^{\deg(f)})^N + N t^{\deg(f)} (1 - t^{\deg(f)})^{N-1} \right) =$$

$$= \lim_{g \to \infty} \prod_{j=1}^{g} ((1 - t^j)^N + N t^j (1 - t^j)^{N-1})^{\varphi_j} = \prod_{j=1}^{\infty} ((1 - t^j)^{N-1} (1 + (N-1)t^j))^{\varphi_j}.$$

$\square$

Computing the asymptotic expansion of this formula leads to the following corollary:

Corollary 6.

The natural density of pairwise coprimeness for N arbitrary polynomials is

$$1 - \binom{N}{2} t + \frac{1}{24} (N-1)(N-2)(3N^2 + 11N) t^2 + O(t^3).$$

Proof.
One has to show

$$\prod_{j=1}^{\infty} ((1 - t^j)^{N-1} (1 + (N-1)t^j))^{\varphi_j}$$

$$= 1 - \binom{N}{2} t + \frac{1}{24} (N-1)(N-2)(3N^2 + 11N) t^2 + O(t^3).$$

One uses the estimations $\varphi_j = \frac{1}{j}t^{-j} + O(t^{-(j-1)})$ as well as

$$F_j := (1 - t^j)^{N-1}(1 + (N-1)t^j) =$$

$$= 1 + \left(\binom{N-1}{2} - (N-1)^2\right)t^{2j} +$$

$$+ \left(\binom{N-1}{2}(N-1) - \binom{N-1}{3}\right)t^{3j} + O(t^{4j}) =$$

$$= 1 + \frac{N-1}{2} \cdot (N - 2 - 2(N-1))t^{2j} +$$

$$+ (N-1)(N-2)\left(\frac{N-1}{2} - \frac{N-3}{6}\right)t^{3j} + O(t^{4j})$$

$$= 1 - \binom{N}{2}t^{2j} + \frac{1}{3}N(N-1)(N-2)t^{3j} + O(t^{4j}).$$

If one chooses x times the term with exponent $-kj$ (for $k \geq 2$) expanding $\prod_{j=1}^{\infty} F_j^{\varphi_j}$, one gets a term of the form $C(N)\binom{\varphi_j}{x}t^{xkj} = O(t^{(k-1)xj})$ with $C(N)$ only depending on N. Thus, one is only interested in the case that $k - 1 = x = j = 1$ and in the case that one number from the set $\{k - 1, x, j\}$ is equal to 2 and the others are equal to 1. In particular, the considered probability is equal to

$$\underbrace{\left(1 - \binom{N}{2}t^2 + \frac{1}{3}N(N-1)(N-2)t^3\right)^{t^{-1}}}_{j=1,\ k\leq 3} \underbrace{\left(1 - \binom{N}{2}t^4\right)^{\frac{1}{2}(t^{-2}-t^{-1})}}_{j=2,\ k\leq 2} + O(t^3)$$

$$= \left(1 - \underbrace{\binom{N}{2}t}_{k=2,\ x=1} + \underbrace{\frac{1}{3}N(N-1)(N-2)t^2}_{k=3,\ x=1} + \underbrace{\binom{t^{-1}}{2}\binom{N}{2}^2 t^4}_{k=2,\ x=2} + O(t^3)\right).$$

$$\cdot \left(1 - \binom{N}{2}t^4 \cdot \frac{1}{2}t^{-2} + O(t^3)\right) + O(t^3)$$

$$= 1 - \binom{N}{2}t + \left(\frac{1}{3}N(N-1)(N-2) + \frac{N^2(N-1)^2}{8} - \frac{N(N-1)}{4}\right)t^2 + O(t^3)$$

$$= 1 - \binom{N}{2}t + \left(\frac{1}{3}N(N-1)(N-2) + \frac{N(N-1)}{8}(N(N-1) - 2)\right)t^2 + O(t^3)$$

$$= 1 - \binom{N}{2}t + \left(\frac{1}{3}N(N-1)(N-2) + \frac{(N-1)(N-2)(N+1)N}{8}\right)t^2 + O(t^3)$$

$$= 1 - \binom{N}{2}t + \frac{1}{24}(N-1)(N-2)(3N^2 + 11N)t^2 + O(t^3),$$

which completes the proof of the corollary. $\qquad\square$

Corollary 6 leads to the same result as Theorem 16 with setting $N_1 = 0$, although different concepts of probability were used. This concordance could be explained in the following way: First, computing the natural density of pairwise coprimeness, those tuples of polynomials which contain a linear polynomial could be neglected. Moreover, the case that $d_i \equiv 0$ for some $i \in \{1, \cdots, n\}$ could be neglected and hence, considering monic polynomials does not change the probability because two polynomials are coprime if and only if the corresponding monic polynomials are coprime. Thus, all degree dependencies of the considered coefficients in the asymptotic expansion could be neglected. Therefore, choosing the polynomials randomly with $\deg(d_i) \leq n_i$, the probability could be regarded as identical for all values $n_i \in \mathbb{N}$ since the set of polynomials with $\deg(d_i) \leq n_i$ is a disjunct union of the sets of polynomials whose degree is a fixed value less or equal to n_i. But the sets defined by the condition $\deg(d_i) \leq n_i$ form a subsequence of $\mathcal{M}_n$. Consequently, if one knows that the limit defining the natural density exists, one could conclude that it is equal to the constant value for this subsequence.

2.4 Counting Mutually Coprime Polynomial Matrices

The aim of this section is to compute the probability that N nonsingular polynomial matrices from $\mathbb{F}[z]^{m \times m}$ are mutually left coprime, which we denote by $P_m(N)$. At first, we consider the probability that two nonsingular polynomial matrices D_1 and D_2 are left coprime. Since $[D_1 \ D_2] \cdot \begin{bmatrix} U_1 & 0 \\ 0 & U_2 \end{bmatrix} = [D_1 U_1 \ D_2 U_2]$, one could assume that both D_1 and D_2 are in Hermite form. Again, we begin with some easily proven bounds before we calculate the asymtotic behaviour.

Theorem 18.
Let $P_m(2)$ be the probability that two nonsingular matrices $D_1, D_2 \in \mathbb{F}[z]^{m \times m}$ (in Hermite form) with $\deg(D_i) = n_i$ for $i = 1, 2$ are left coprime. Then,

$$1 - m(n_1 + n_2)t \leq P_m(2) \leq 1 - t^{2m-1}.$$

Proof.
The proof could be done completely analogue to the proof of Theorem 11 since D_1 and D_2 are left coprime if and only if $\begin{pmatrix} D_1^\top \\ D_2^\top \end{pmatrix}$ is right prime. For the lower bound, the only difference is that here the minors $D_1^\top$ and $D_2\top$ are of degrees n_1 and n_2, respectively. □

Again, this bound is not very sharp, as we will see in the following.

Theorem 19.
The probability that two matrices $D_1, D_2 \in \mathbb{F}[z]^{m \times m}$ in Hermite form with $\deg(D_i) = n_i$ for $i = 1, 2$ are left coprime is equal to $1 - t^m + O(t^{m+1})$.

Proof.
Since the statement is already known for $m = 1$ (see Lemma 5), in the following, it is assumed that $m \geq 2$. At first, it should be shown that the cardinality of

$S \subset X := X(n_1, n_2)$ of matrices for which $\mathcal{D}_2$ is not left prime is $O(|X| \cdot t^m)$ and that the cardinality of the subset of S for which $\mathcal{D}_2$ is not of simple form is $O(|X| \cdot t^{m+1})$. Denote the entries of $\mathcal{D}_2$ by $\mathfrak{d}_{ij}$ for $i = 1, \ldots, m$ and $j = 1, \ldots, 2m$ and choose $z_* \in \mathbb{F}$ such that $\mathcal{D}_2(z_*)$ is not of full row rank. As in the method of iterated row operations (see Lemma 9 (b)), start considering the first row of this matrix. Either it is identically zero, which means $d_{11}^{(1)}(z_*) = d_{11}^{(2)}(z_*) = 0$, or one could assume without restriction that $d_{11}^{(2)}(z_*) \neq 0$. For the first case, one has a cardinality of $|X| \cdot t$ due to the fact that the two polynomials have a common zero. Moreover, they cannot be constant, i.e. $\kappa_m^{(i)} \geq 1$ for $i = 1, 2$. Thus, it follows from (2.1) that one has an additional factor for the cardinality of at most $t^{2(m-1)}$, which in summary, leads to a cardinality of $O(|X| \cdot t^{2m-1}) = O(|X| \cdot t^{m+1})$ for $m \geq 2$ and one is finished. If $d_{11}^{(2)}(z_*) \neq 0$, one proceeds as in the method of iterated row operations, i.e. in the first step, one subtracts multiples of the first row to the rows further down in such way that all entries in column $m+1$ but $d_{11}^{(2)}(z_*)$ are nullified. From Lemma 9 (b), one knows that there exist $k \in \{1, \ldots, m\}$, a set of column indices $\{j_1, \ldots, j_{k-1}\} \subset \{1, \ldots, 2m\}$ and values $\lambda_r \in \mathbb{F}(z_*)$, which (only) depend on entries $\mathfrak{d}_{ij}$ of $\mathcal{D}_2$ with $j \in \{j_1, \ldots, j_{k-1}\}$ and on z_*, such that

$$\mathfrak{d}_{kj}(z_*) = \sum_{r=1}^{k-1} \mathfrak{d}_{rj}(z_*) \cdot \lambda_r \quad \text{for } j \in \{1, \ldots, 2m\} \setminus \{j_1, \ldots, j_{k-1}\}. \tag{2.18}$$

Moreover, since D_1 and D_2 are lower triangular, it holds $\mathfrak{d}_{ij} \equiv 0$ for $i < j \leq m$ or $i + m < j \leq 2m$. Therefore, (2.18) is equivalent to

$$\mathfrak{d}_{kj}(z_*) = \sum_{r=1}^{k-1} \mathfrak{d}_{rj}(z_*) \cdot \lambda_r \text{ for } j \in \{1, \ldots, k, m+1, \ldots, m+k\} \setminus \{j_1, \ldots, j_{k-1}\}. \tag{2.19}$$

Note that $\{j_1, \ldots, j_{k-1}\}$ is a subset of $\{1, \ldots, k-1, m+1, \ldots, m+k-1\}$ because $\mathfrak{d}_{i,j_i}(z_*) \neq 0$ for $1 \leq i \leq k-1$ per construction and hence, $j_i \leq i \leq k-1$ or $m + k - 1 \geq m + i \geq j_i > m$. Furthermore, it follows that $d_{kk}^{(1)}(z_*) = \mathfrak{d}_{kk}(z_*) = \sum_{r=1}^{k-1} \mathfrak{d}_{rk}(z_*) \cdot \lambda_r = 0$ and $d_{kk}^{(2)}(z_*) = \mathfrak{d}_{k,m+k}(z_*) = \sum_{r=1}^{k-1} \mathfrak{d}_{r,m+k}(z_*) \cdot \lambda_r = 0$. This could also be seen directly by observing that the iteration process only changes entries beyond the diagonals of the matrices D_1 and D_2.

Thus, one has the conditions $d_{kk}^{(1)}(z_*) = d_{kk}^{(2)}(z_*) = 0$, which moreover, ensure $\kappa_{m-k+1}^{(i)} \geq 1$ for $i = 1, 2$. In particular, this implies that the polynomials $d_{kj}^{(i)}$ for $j < k$ and $i = 1, 2$ are not fixed to zero by degree restrictions. But since $d_{kj}^{(1)} = \mathfrak{d}_{kj}$ and $d_{kj}^{(2)} = \mathfrak{d}_{k,j+m}$ for $j = 1, \ldots, k-1$, one knows from (3.8) that $2(k-1) - (k-1) = k-1$ of these polynomials are fixed at z_* by the remaing polynomials of $\mathcal{D}_2$. We fix $g := g_{z_*}$ and apply Lemma 7 (a) and (b) with $\tilde{w} \equiv 1$ and $w_j := \sum_{r=1}^{k-1} \mathfrak{d}_{rj} \cdot \lambda_r$ for $j \in \{1, \ldots, k-1, m+1, \ldots, m+k-1\} \setminus \{j_1, \ldots, j_{k-1}\}$. One obtains a cardinality that is $O(|X| \cdot \varphi_g \cdot t^{2g+k-1}) = O(|X| \cdot t^{g+k-1})$ for the above conditions. Additionally, one gets the factor $t^{2(m-k)}$ from (2.1) since $\kappa_{m-k+1}^{(i)} \geq 1$. In summary, the cardinality is $O(|X| \cdot t^{2m-k+g-1}) = O(|X| \cdot t^{m+1})$ for $k \leq m - 1$. For $k = m$, one has a factor of $O(|X| \cdot t^m)$. If there is no simple form, it follows from (2.1) that this cardinaliy is decreased by a factor of at most t. Hence, the overall cardinality is $O(|X| \cdot t^{m+1})$. This shows the claim of the first paragraph of this proof for the case that g is fixed. But since g is bounded above by $\min(n_1, n_2)$, it is also valid for summing up over all possible values for g. Moreover, note that the considered cardinality is $O(|X| \cdot t^{m+g-1}) = O(|X| \cdot t^{m+1})$ for $g \geq 2$, even for simple form.

It remains to compute the coefficient of t^m. It follows from the previous paragraph that for this computation it is sufficient to consider only matrices of the form

$$D_j = \begin{bmatrix} I_{m-1} & 0 \\ d_1^{(j)} \cdots d_{m-1}^{(j)} & d_m^{(j)} \end{bmatrix}$$

for $j = 1, 2$ for which there exists $z_* \in \mathbb{F}$ such that $\begin{bmatrix} D_1(z_*) & D_2(z_*) \end{bmatrix}$ is singular, which is the case if and only if $d_m^{(1)}(z_*) = d_m^{(2)}(z_*) = 0$ and $d_k^{(1)}(z_*) = d_k^{(2)}(z_*)$ for $1 \leq k \leq m - 1$. According to Lemma 7 (a) and (b) with $\tilde{w} \equiv 1$, $w_j = d_j^{(2)}$ for $j = 1, \ldots m - 1$ and $g = 1$, the probability for this is equal to $\varphi_1 \cdot t^{2+m-1} = t^m$. Hence, the proof of the theorem is complete. $\square$

Comparison of the preceding result with the formula for the natural density from [17], leads to similar observations as in Section 2.2. The natural density of left primeness for $\mathcal{D} \in \mathbb{F}[z]^{m \times 2m}$ is equal to $\prod_{j=m}^{2m-1}(1 - t^j)$. This coincides with our asymptotic result for the uniform probability. Moreover, the asymptotic formula does not depend on any degrees. But actually, the exact probability depends on the degrees of the determinants of the constituent matrices.

Example 5.
The probability that $[D_1 \, D_2] \in \mathbb{F}[z]^{2 \times 4}$ with $\deg(\det(D_i)) = 1$ for $i = 1, 2$ is left prime is equal to $1 - \frac{1}{t^{-2}+t^{-1}} = 1 - t^2 \sum_{k=0}^{\infty} (-t)^k > (1 - t^2)(1 - t^3)$.

Proof.
Again we can assume that both D_1 and D_2 are in Hermite form. Thus, possible structures are

$$D_i = \begin{bmatrix} z - a_i & 0 \\ 0 & 1 \end{bmatrix} \text{ and } D_i = \begin{bmatrix} 1 & 0 \\ b_i & z - c_i \end{bmatrix} \text{ with } a_i, b_i, c_i \in \mathbb{F} \text{ for } i = 1, 2.$$

If D_1 and D_2 have different structures, it is obvious that $[D_1 \, D_2]$ is left prime. If both matrices are of the first structure, they are left coprime if and only if $a_1 \neq a_2$. And finally, if both matrices are of the second structure, they are not left coprime if and only if $b_1 = b_2$ and $c_1 = c_2$. In summary, the two matrices are left coprime if and only if they are not identical. Since there are $t^{-2} + t^{-1}$ possibilities for each matrix, the probability of left coprimeness is $1 - \frac{t^{-2}+t^{-1}}{(t^{-2}+t^{-1})^2} = 1 - \frac{1}{t^{-2}+t^{-1}}$.

The series expansion of this term is obtained by the formula for the geometric series. But we do not employ the series expansion to show the lower bound. It is derived by the simple computation

$$(1-t^2)(1-t^3) = 1 - \frac{(t^2+t^3-t^6)(t^{-2}+t^{-1})}{t^{-2}+t^{-1}} = 1 - \frac{1+2t+t^2-t^4-t^5}{t^{-2}+t^{-1}}$$

since $2t + t^2 > t^4 + t^5$. □

Analogous to the considerations of Section 2.2, one reason why the probability of the preceding example differs from the natural density is that some of the entries are fixed ones. Since these entries could never be zero, they lead to a higher probability for left primeness. The following example considers Hermite forms with no fixed ones.

Example 6.
For $m \geq 2$, the probability that $[D_1 \ D_2] \in \mathbb{F}[z]^{m \times 2m}$, where D_1 and D_2 are in Hermite form and have no diagonal entries equal to one, is left prime is upper bounded by

$$\prod_{j=1}^{m} 1 - t^{2j-1} < \prod_{j=m}^{2m-1} 1 - t^j.$$

Proof.
For the left primeness of the matrix it is necessary that the polynomials of each row are coprime. The formula follows from the fact that row j contains $2j$ polynomials which are not fixed zeros by using Corollary 2. □

One reason why the probability of the preceding example differs from the formula for the natural density is that some of the entries are fixed zeros due to the lower triangular form of the two constituent matrices. They lead to a lower probability for left primeness. Since matrices in Hermite form always contain fixed zeros, the following example considers the easiest possible Kronecker-Hermite form which has no fixed ones and therefore no fixed zeros, too.

Example 7.
For $t = \frac{1}{2}$, the probability that $\begin{bmatrix} z+a_1 & a_3 & z+b_1 & b_3 \\ a_4 & z+a_2 & b_4 & z+b_2 \end{bmatrix}$ with $a_i, b_i \in \mathbb{F}$ is left prime is equal to $(1 - 2^{-2})(1 - 2^{-3}) = \frac{168}{256}$.

Proof.
With the help of matlab, one gets $\frac{170}{256}$ for the probability that the matrix has full rank for $z = 0$ and $z = 1$. It remains to count the cases for which $z^2 + z + 1$ divides the six full size minors:

$$\begin{array}{lll}
z^2 + (a_1+a_2)z + a_1a_2 + a_3a_4 = z^2 + z + 1 & \Leftrightarrow & a_2 \neq a_1, \ a_3 = a_4 = 1 \\
z^2 + (a_1+b_2)z + a_1b_2 + a_4b_3 = z^2 + z + 1 & \Leftrightarrow & b_2 \neq a_1, \ a_4 = b_3 = 1 \\
z^2 + (a_2+b_1)z + a_2b_1 + a_3b_4 = z^2 + z + 1 & \Leftrightarrow & a_2 \neq b_1, \ a_3 = b_4 = 1 \\
z^2 + (b_1+b_2)z + b_1b_2 + b_4b_3 = z^2 + z + 1 & \Leftrightarrow & b_2 \neq b_1, \ b_4 = b_3 = 1 \\
(b_4+a_4)z + a_1b_4 + a_4b_1 & & \\
(b_3+a_3)z + a_2b_3 + a_3b_2 & &
\end{array}$$

To be divided by z^2+z+1, the minors in lines one to four have to be equal to z^2+z+1. This is the case if and only if $a_1 = b_1 \neq a_2 = b_2$ and $a_3 = a_4 = b_3 = b_4 = 1$. One easiliy verifies that the minors in the last two lines are equal to the zero polynomial in these cases and hence, are divided by $z^2 + z + 1$, too. Consequently, there are two possibilities for the choice of the parameters and one gets $\frac{170}{256} - \frac{2}{256} = \frac{168}{256}$ for the overall probability. $\qquad\square$

As in Section 2.2, the preceding observations lead to the conjecture that the natural density computed in [17] coincides with the probability for matrices in Kronecker-Hermite from with fixed column degrees unequal to zero.

In the following, we want to extend the previous results to $N \geq 3$ matrices. But before we approach our actual goal, which is to compute the probability that N matrices are mutually left coprime, we first consider the case of pairwise left coprimeness, which could be deduced from the case $N = 2$, where pairwise left and mutual left coprimeness coincide.

Theorem 20.
For $m \geq 1$, the probability of N matrices $D_i \in \mathbb{F}[z]^{m \times m}$ in Hermite form with $\deg(\det(D_i)) = n_i$ for $i = 1, \ldots, N$ to be pairwise left coprime is equal to $1 - \frac{N(N-1)}{2} t^m + O(t^{m+1})$.

Proof.
Let S be the subset of $X := X(n_1, \ldots, n_N)$ (see Definition 11) for which the tuples consist of pairwise left coprime matrices and $\mathcal{E} := \{ij \mid 1 \leq i < j \leq N\}$. Thus, $S = X \setminus \bigcup_{r \in R} S_r$ with $R = \mathbb{F} \times \mathcal{E}$ and $S_{(z_*, ij)} = \{(D_1, \ldots, D_N) \subset X \mid [\; D_i(z_*) \quad D_j(z_*) \;]$ is singular$\}$. By the inclusion-exclusion-principle, one obtains:

$$|S| = \sum_{T \subset R} (-1)^{|T|} |S_T| \quad \text{with} \quad S_T = \bigcap_{r \in T} S_r \quad \text{and} \quad S_\emptyset = X.$$

From Theorem 19, it follows that the probability $\frac{|S|}{|X|}$ is equal to $1 + O(t^m)$. Moreover, from the proof of Theorem 19, it follows that for the computation of the coefficient of t^m, it is sufficient to consider only matrices of the form

$$D_j = \begin{bmatrix} I_{m-1} & 0 \\ d_1^{(j)} \ldots d_{m-1}^{(j)} & d_m^{(j)} \end{bmatrix}$$

for $j = 1, \ldots, N$ for which there exist $z_* \in \mathbb{F}$ and $1 \leq i < j \leq N$ such that $[\; D_i(z_*) \quad D_j(z_*) \;]$ is singular. Recall that $[\; D_i(z_*) \quad D_j(z_*) \;]$ is singular if and only if $d_m^{(i)}(z_*) = d_m^{(j)}(z_*) = 0$ and $d_k^{(i)}(z_*) = d_k^{(j)}(z_*)$ for $1 \leq k \leq m - 1$.
For the case that there exist $\tilde{z}_* \in \mathbb{F}$ and $1 \leq u < v \leq N$ with $(u, v) \neq (i, j)$ such that $[\; D_u(\tilde{z}_*) \quad D_v(\tilde{z}_*) \;]$ is singular, too, it is obvious that the probability is $O(t^{2m}) = O(t^{m+1})$ if $\{i, j\} \cap \{u, v\} = \emptyset$. If $\{i, j\} \cap \{u, v\} \neq \emptyset$, assume without restriction that $j = u$. Then, one could choose $d_1^{(j)}, \ldots, d_{m-1}^{(j)}$ as well as $z_*, \tilde{z}_* \in \mathbb{F}$ arbitrarily, which affects that $d_1^{(i)}, \ldots, d_{m-1}^{(i)}$ are fixed at z_* and $d_1^{(v)}, \ldots, d_{m-1}^{(v)}$ are fixed at $\tilde{z}_*$. If $\tilde{z}_* = z_*$, it follows from Lemma 7 (a) and (b) that the probability is

$t^{-1+3+2(m-1)} = O(t^{m+1})$. If $\tilde{z}_* \neq z_*$, for which there are $O(t^{-2})$ possibilities, Lemma 7 (a) and (b) lead to a probability of $O(t^{-2+2+2+2(m-1)}) = O(t^{m+1})$.

Thus, in all these cases, the probability is $O(t^{m+1})$, which means that they do not contribute anything to the coefficient of t^m. Consequently, only $T \subset R$ of the form $T = \{(z_*, ij)\}$ with $z_* \in \mathbb{F}$ give a contribution to the coefficient of t^m, namely $|S_T| = |X| \cdot t^m$ (see end of proof of Theorem 19). This leads to

$$\frac{|S|}{|X|} = 1 - \sum_{z_* \in \mathbb{F}, ij \in \mathcal{E}} \frac{|S_{(z_*,ij)}|}{|X|} + O(t^{m+1}) = 1 - \sum_{ij \in \mathcal{E}} t^m + O(t^{m+1}) =$$

$$= 1 - \frac{N(N-1)}{2} t^m + O(t^{m+1}).$$

□

Remark 10.

For $m \geq 2$, $N \geq 3$ mutual left coprimeness is a stronger condition than pairwise left coprimeness, as the following example shows.

Example 8.

Consider the pairwise left coprime matrices in Hermite form

$$D_1(z) = \begin{bmatrix} 1 & 0 \\ 1 & z \end{bmatrix}, \ D_2(z) = \begin{bmatrix} 1 & 0 \\ 0 & z \end{bmatrix} \text{ and } D_3(z) = \begin{bmatrix} z & 0 \\ 0 & 1 \end{bmatrix}. \text{ We show in two}$$

different ways that they are, however, not mutually left coprime. One way to see that is to consider

$$\begin{bmatrix} D_1(z) & D_2(z) & 0 \\ 0 & D_2(z) & D_3(z) \end{bmatrix} = \begin{bmatrix} 1 & 0 & 1 & 0 & 0 & 0 \\ 1 & z & 0 & z & 0 & 0 \\ 0 & 0 & 1 & 0 & z & 0 \\ 0 & 0 & 0 & z & 0 & 1 \end{bmatrix}.$$

Since this matrix is singular for $z = 0$, D_1, D_2 and D_3 are not mutually left coprime. A second way to show this is to compute a least common right multiple of e.g. D_2 and D_3, denoted by D_{23}. It is easy to see that one can choose $D_{23}(z) = \begin{bmatrix} z & 0 \\ 0 & z \end{bmatrix}$, which clearly is not left coprime with $D_1(z)$. The following remark shows in particular, that for $N = 3$ only matrices whose determinants have a common zero (here $z = 0$) could be pairwise but not mutually left coprime.

To calculate the probability of mutual left coprimeness, we will firstly prove a recursion formula for it. Therefore, we have to investigate what happens if $N - 1$ of the considered matrices are mutually left coprime but all N matrices together do not have this property. This is done by the following remark.

Remark 11.

Let $D_1, \ldots, D_N$ be not mutually left coprime with $\xi D_N(z_*) = 0$ for some $z_* \in \overline{\mathbb{F}}$ and $\xi \in \mathbb{F}(z_*)^{m(N-1)}$. Moreover, every set consisting of $N - 1$ of these matrices should be mutually left coprime. Then, it holds $\xi \in (\mathbb{F}(z_*) \setminus \{0\})^{m(N-1)}$ and $\det(D_i(z_*)) = 0$ for $i = 1, \ldots, N$.

Proof.
According to Theorem 6, $D_1, \ldots, D_N$ are mutually left coprime if and only if

$$\mathcal{D}_N := \begin{bmatrix} D_1 & D_2 & 0 & 0 \\ 0 & \ddots & \ddots & 0 \\ 0 & 0 & D_{N-1} & D_N \end{bmatrix} \text{ is left prime.}$$

Since this is not true, there exist $z_* \in \overline{\mathbb{F}}$ and $\xi := (\xi_1, \ldots, \xi_{N-1}) \neq 0$ with $\xi_i \in \mathbb{F}(z_*)^{1 \times m}$ for $i = 1, \ldots, N-1$ such that $\xi \mathcal{D}_N(z_*) = 0$, i.e.
$\xi_1 D_1(z_*) = 0$, $(\xi_{i-1} + \xi_i) D_i(z_*) = 0$ for $i = 2, \ldots, N-1$ and $\xi_{N-1} D_N(z_*) = 0$.
Consequently, one has to show
$\xi_1 \neq 0$, $\xi_{i-1} + \xi_i \neq 0$ for $i = 2, \ldots, N-1$ and $\xi_{N-1} \neq 0$
if every proper subset of $\{D_1, \ldots, D_N\}$ consists of mutually left coprime matrices. This is shown per contradiction.
If $\xi_{N-1} = 0$, i.e. $\tilde{\xi} := (\xi_1, \ldots, \xi_{N-2}) \neq 0$, it follows $\tilde{\xi} \mathcal{D}_{N-1}(z_*) = 0$, i.e. $D_1, \ldots, D_{N-1}$ would not be mutually left coprime. Similarly, if $\xi_1 = 0$,
$D_2, \ldots, D_N$ would not be mutually left coprime. To show $\xi_{i-1} + \xi_i \neq 0$ for $i = 2, \ldots, N-1$, one needs $\xi_i \neq 0$ for $i = 2, \ldots, N-2$. If $\xi_k = 0$ for some $k = 2, \ldots, N-2$, it follows $(\xi_1, \ldots, \xi_{k-1}) \neq 0$ or $(\xi_{k+1}, \ldots, \xi_{N-1}) \neq 0$. In the first case, $D_1, \ldots, D_k$ would not be mutually left coprime, in the second case, $D_{k+1}, \ldots, D_N$ would not be mutually left coprime. Now, assume $\xi_{k-1} + \xi_k = 0$ for some $k = 2, \ldots, N-1$. Define $\hat{\xi} := (\hat{\xi}_1, \ldots, \hat{\xi}_{N-2})$ with $\hat{\xi}_i = \xi_i$ for $i \leq k-1$ and $\hat{\xi}_i = -\xi_{i+1}$ for $i \geq k$. Then holds $\hat{\xi} \neq 0$ and $\hat{\xi}_1 D_1(z_*) = 0$, $(\hat{\xi}_{i-1} + \hat{\xi}_i) D_i(z_*) = 0$ for $i = 2, \ldots, k-1$, $(\hat{\xi}_{i-1} + \hat{\xi}_i)(-D_{i+1}(z_*)) = (\xi_i + \xi_{i+1}) D_{i+1}(z_*) = 0$ for $i = k, \ldots, N-2$ and $\hat{\xi}_{N-2}(-D_N(z_*)) = \xi_{N-1} D_N(z_*) = 0$. This means that $D_1, \ldots, D_{k-1}, -D_{k+1}, \ldots, -D_N$ are not mutually left coprime. But then $D_1, \ldots, D_{k-1}, D_{k+1}, \ldots, D_N$ are not mutually left coprime, too. To see that let $\mathcal{D}_{N \backslash k}^-$ be constructed out of $D_1, \ldots, D_{k-1}, -D_{k+1}, \ldots, -D_N$ and $\mathcal{D}_{N \backslash k}$ out of $D_1, \ldots, D_{k-1}, D_{k+1}, \ldots, D_N$. Because of

$$\mathcal{D}_{N \backslash k} = \mathcal{D}_{N \backslash k}^- \begin{bmatrix} I_{m(k-1)} & 0 \\ 0 & -I_{m(N-k)} \end{bmatrix}, \text{ left primeness of } \mathcal{D}_{N \backslash k}^- \text{ is equivalent to left}$$

primeness of $\mathcal{D}_{N \backslash k}$. Consequently, the proof is complete. $\qquad \square$

With a similar reasoning, one could show a sufficient criterion for mutual left coprimeness, as done in the following.

Remark 12.
If $\det(D_1), \ldots, \det(D_N)$ are pairwise coprime, $D_1, \ldots, D_N$ are mutually left coprime.

Proof.
We assume that $D_1, \ldots, D_N$ are not mutually left coprime and show per induction with respect to N that then $\det(D_1), \ldots, \det(D_N)$ are not pairwise coprime. For $N = 2$, D_1 and D_2 are not left coprime if and only if there exists $z_* \in \overline{\mathbb{F}}$ such that $[D_1(z_*) \; D_2(z_*)]$ has no full row rank. Clearly, this implies that $\det(D_1(z_*)) = \det(D_2(z_*)) = 0$ and therefore, $\det(D_1)$ and $\det(D_2)$ are not coprime.
Now let $D_1, \ldots, D_N$ with $N \geq 3$ be not mutually left coprime. As in the preceding proof, one knows that there exist $z_* \in \overline{\mathbb{F}}$ and $\xi := (\xi_1, \ldots, \xi_{N-1}) \neq 0$ with $\xi_i \in \overline{\mathbb{F}}^{1 \times m}$ for $i = 1, \ldots, N-1$ such that $\xi_1 D_1(z_*) = 0$, $(\xi_{i-1} + \xi_i) D_i(z_*) = 0$ for $i = 2, \ldots, N-1$

and $\xi_{N-1}D_N(z_*) = 0$. If $\xi_1 = 0$, $D_2, \ldots, D_N$ would be not mutually left coprime. But then, one knows that $\det(D_2), \ldots, \det(D_N)$ are not pairwise coprime per induction and hence, $\det(D_1), \ldots, \det(D_N)$ are not pairwise coprime, too. Similarly, if $\xi_{N-1} = 0$, $D_1, \ldots, D_{N-1}$ are not mutually left coprime and thus, $\det(D_1), \ldots, \det(D_{N-1})$ not pairwise coprime. Consequently, one could assume $\xi_1 \neq 0$ and $\xi_{N-1} \neq 0$. But since $\xi_1 D_1(z_*) = 0 = \xi_{N-1}D_N(z_*)$, this implies that $\det(D_1(z_*)) = \det(D_N(z_*)) = 0$, which means that $\det(D_1), \ldots, \det(D_N)$ are not pairwise coprime. □

However, this criterion for mutual left coprimeness is far away from being necessary. Consider for example the matrices $\begin{bmatrix} 1 & 0 \\ 0 & z \end{bmatrix}$, $\begin{bmatrix} z & 0 \\ 0 & 1 \end{bmatrix}$ and $\begin{bmatrix} 1 & 0 \\ 0 & 1 \end{bmatrix}$, which are clearly mutually left coprime but the determinants of the first two matrices have the common zero 0 and are thus not coprime.

The following theorem represents the crucial step in computing the probability of mutual left coprimeness by providing a recursion formula for it.

Theorem 21.
For $N \geq 2$, the probability that N matrices $D_i \in \mathbb{F}[z]^{m\times m}$ in Hermite form with $\deg(\det(D_i)) = n_i$ for $i = 1, \ldots, N$ are mutually left coprime is equal to

$$P_m(N) = 1 + \sum_{k=1}^{N-2}(-1)^k \binom{N}{k}(1 - P_m(N-k)) - \sum_{i=N-1}^{\min(m,N-1)} t^m + O(t^{m+1}),$$

where $P_m(N-k)$ denotes the probability that $N-k$ such matrices are mutually left coprime.

Proof.
For $N = 2$, the formula has already been proven in Theorem 19. Therefore, one could assume $N \geq 3$. Let $mut(N)$ be the subset of $X(N) := X(n_1, \ldots, n_N)$ for which the tuples consist of mutually left coprime matrices. Moreover, for $i = 1, \ldots, N$, $A_i(N)$ should denote the subset of $X(N)$ for which the matrices in the set $\{D_1, \ldots, D_N\} \setminus \{D_i\}$ are not mutually left coprime. Finally, define $A_{N+1}(N) := (X(N) \setminus mut(N)) \cap (X(N) \setminus \bigcup_{i=1}^{N} A_i(N))$, i.e. $A_{N+1}(N)$ consists of those tuples that are not mutually left coprime but all subsets of $N-1$ matrices are mutually left coprime.
Thus, $mut(N) = X(N) \setminus \bigcup_{i=1}^{N+1} A_i(N)$. By the inclusion-exclusion-principle, one obtains:

$$|mut(N)| = \sum_{I \subset \{1,\ldots,N+1\}} (-1)^{|I|}|A_I(N)|$$

$$\text{with} \quad A_I(N) = \bigcap_{i \in I} A_i(N) \quad \text{and} \quad A_\emptyset(N) = X(N).$$

From the definition of the $A_i(N)$, it follows $A_{N+1}(N) \cap A_i(N) = \emptyset$ for $i = 1, \ldots, N$

and consequently,

$$|mut(N)| = |X(N)| - |A_{N+1}(N)| + \sum_{\emptyset \neq I \subset \{1,\ldots,N\}} (-1)^{|I|} |A_I(N)|. \tag{2.20}$$

In the following, it is used that $D_1, \ldots, D_N$ are mutually left coprime if and only if

$$\mathcal{D}_N := \begin{bmatrix} D_1 & D_2 & 0 & 0 \\ 0 & \ddots & \ddots & 0 \\ 0 & 0 & D_{N-1} & D_N \end{bmatrix} \text{ is left prime; see Theorem 6.}$$

At first, it is shown that the cardinality of $X(N) \setminus mut(N)$ is $O(|X(N)| \cdot t^m)$ and that the cardinality of the subset of $X(N) \setminus mut(N)$ which contains only tuples of matrices such that $\mathcal{D}_N$ is not of simple form is $O(|X(N)| \cdot t^{m+1})$. Doing this, one uses the following claim.

Claim 1:

If $D_i = \begin{bmatrix} v_i & 0 \\ w_i & D_i^{(m-1)} \end{bmatrix}$ with $v_i \in \mathbb{F}[z]$ and $v_i(z_*) \neq 0$ for $i = 1, \ldots, N$ and some $z_* \in \overline{\mathbb{F}}$, it holds

$$\mathrm{rk}(\mathcal{D}_N(z_*)) =$$

$$= \mathrm{rk}\left(\begin{bmatrix} w_1 - \frac{v_1}{v_2}w_2 & D_1^{(m-1)} & D_2^{(m-1)} & & \\ (\frac{w_3}{v_3} - \frac{w_2}{v_2})v_1 & D_2^{(m-1)} & D_3^{(m-1)} & & \\ \vdots & & \ddots & \ddots & \\ (-1)^N (\frac{w_{N-1}}{v_{N-1}} - \frac{w_N}{v_N})v_1 & & & D_{N-1}^{(m-1)} & D_N^{(m-1)} \end{bmatrix}(z_*) \right) +$$

$$+ N - 1.$$

Proof of claim 1:
This proof uses a method that is similar to the method of iterated row operations introduced in Lemma 9 (b). However, one applies only one iteration step to special subblocks of $\mathcal{D}_N(z_*)$.
One starts adding row $(N-2)m+1$ of $\mathcal{D}_N(z_*)$ times $\frac{-w_{N,r}}{v_N}(z_*)$ to row $(N-2)m+1+r$ for $r = 1, \ldots, m-1$. Here, $w_{N,r}$ denotes the r-th component of the vector w_N. Afterwards, deleting row $(N-2)m+1$ and column $(N-1)m+1$ decreases the rank of $\mathcal{D}_N(z_*)$ by 1. This affects only the block $[D_{N-1} \; D_N](z_*)$, whose first row and $(m+1)$-th column are deleted and whose first column is changed to $(w_{N-1} - \frac{v_{N-1}}{v_N}w_N)(z_*)$. Moreover, some zeros of the zero blocks are deleted. Now, one continues adding multiples of row $(N-3)m+1$ to all rows further down in such way that the entries in these rows which are in column $(N-2)m+1$ are nullified. This additionally changes the entries of these rows that are in column $(N-3)m+1$ but no other entries. Afterwards, row $(N-3)m+1$ and column $(N-2)m+1$ are deleted decreasing the

rank by 1. Hence, per induction with respect to N, one could assume

$$\mathrm{rk}(\mathcal{D}_N(z_*)) =$$

$$\mathrm{rk}\left(\left[\begin{array}{cccccc} v_1 & 0 & v_2 & & 0 & \\ w_1 & D_1^{(m-1)} & w_2 & & D_2^{(m-1)} & \\ & & w_2 - \frac{v_2}{v_3}w_3 & & D_2^{(m-1)} & D_3^{(m-1)} \\ & & \vdots & & & \ddots & \ddots \\ & & (-1)^{N-1}\left(\frac{w_{N-1}}{v_{N-1}} - \frac{w_N}{v_N}\right)v_2 & & D_{N-1}^{(m-1)} & D_N^{(m-1)} \end{array}\right](z_*)\right)$$

$$+ N - 2.$$

Now, one adds the first row to all rows beyond in such way that column $m+1$ is nullified and afterwards, deletes the first row and $(m+1)$-th column. Doing this, one gets

$$\mathrm{rk}(\mathcal{D}_N(z_*)) =$$

$$\mathrm{rk}\left(\left[\begin{array}{cccccc} w_1 - \frac{v_1}{v_2}w_2 & & D_1^{(m-1)} & D_2^{(m-1)} & & \\ \left(\frac{w_3}{v_3} - \frac{w_2}{v_2}\right)v_1 & & D_2^{(m-1)} & D_3^{(m-1)} & & \\ \vdots & & & \ddots & \ddots \\ (-1)^N\left(\frac{w_{N-1}}{v_{N-1}} - \frac{w_N}{v_N}\right)v_1 & & & & D_{N-1}^{(m-1)} & D_N^{(m-1)} \end{array}\right](z_*)\right)$$

$$+ N - 1$$

and claim 1 is proven.

Denote by $\mathcal{D}_N^{(\tilde{m})}$ the matrix formed by blocks which consist of the last $\tilde{m}$ columns and rows of the matrices D_i for $i = 1, \ldots, N$ and define the sets

$$A(\tilde{m}, k) := \{\mathcal{D}_N \text{ with } \mathrm{rk}(\mathcal{D}_N^{(\tilde{m})})(z_*) \leq (N-1)\tilde{m} - k \text{ for some } z_* \in \bar{\mathbb{F}}\} \text{ and}$$
$$A^f(\tilde{m}, k) := A(\tilde{m}, k) \cap \{\mathcal{D}_N \text{ with no simple form}\}.$$

Claim 2:
For $\tilde{m}, k \in \mathbb{N}$ with $\tilde{m} + k \leq m + 1$, the cardinality of $A(\tilde{m}, k)$ is $O(|X(N)| \cdot t^{\tilde{m}+k-1})$ and the cardinality of $A^f(\tilde{m}, k)$ is $O(|X(N)| \cdot t^{\tilde{m}+k})$.

Proof of claim 2:
The proof is done per induction with respect to $\tilde{m}$ and starts with the base clause $\tilde{m} = 1$. For $k > N - 1$, it holds $A(1, k) = A^f(1, k) = \emptyset$, which has cardinality $O(|(X(N)| \cdot t^{m(n_1 + \cdots + n_N)}) = O(|X(N)| \cdot t^{\tilde{m}+k+1})$ since $m(n_1 + \cdots + n_N) \geq mN \geq (\tilde{m} + k - 1) \cdot 2 = 2k \geq k + 2 = \tilde{m} + k + 1$ because $k \geq N \geq 2$. Thus, it is sufficient to consider the case $k \leq N - 1$. The blocks that form $\mathcal{D}_N^{(1)}$ are just the scalar polynomials $d^{(i)} := d_{m,m}^{(i)}$ for $i = 1, \ldots, N$. In $A(1, k)$, there exists $z_* \in \bar{\mathbb{F}}$ such that all the matrices consisting of a subset of $N - k$ rows of $\mathcal{D}_N^{(1)}$ are not of full rank at z_*. Especially, the first $N - k$ rows are linearly dependent at z_*, which is equivalent to the fact that

at least two of the polynomials $d^{(1)}, \ldots, d^{(N-k+1)}$ are zero at z_*. Since permutating the set $\{d^{(1)}, \ldots, d^{(N)}\}$ does not change the rank of $\mathcal{D}_N^{(1)}$, every subset of $N - k + 1$ polynomials contains two polynomials that are zero at z_*.

Now, one proceeds in the following way: First, choose the polynomials $d^{(1)}, \ldots, d^{(N-k+1)}$ and denote the polynomials that are zero at z_* by $d_{1,1}$ and $d_{1,2}$. Then, consider the set of polynomials $\{d^{(1)}, \ldots, d^{(N-k+2)}\} \setminus \{d_{1,1}\}$ and iterate this procedure until ending up with the set $\{d^{(1)}, \ldots, d^{(N)}\} \setminus \{d_{1,1}, \ldots, d_{k-1,1}\}$. In summary, at least the $k + 1$ different polynomials $d_{1,1}, \ldots, d_{k-1,1}, d_{k,1}, d_{k,2}$ are zero at z_*. Hence, the cardinality of $A(1, k)$ is $O(|X(N)| \cdot t^{1+k-1})$ (see Lemma 5). Moreover, the cardinality of $A^f(1, k)$ is $O(|X(N)| \cdot t^{k+1})$ since the not simple form decreases the cardinality by at least the factor t; see (2.1).

For the step from $\tilde{m}$ to $\tilde{m} + 1$, three cases are distinguished.

Case 1: $k > (N - 1)(\tilde{m} + 1)$

Here, $A(\tilde{m} + 1, k) = A^f(\tilde{m} + 1, k) = \emptyset$, which has a cardinality that is $O(|X(N)| \cdot t^{m(n_1 + \cdots + n_N)}) = O(|X(N)| \cdot t^{\tilde{m}+k+1})$ since $m(n_1 + \cdots + n_N) \geq 2m \geq m+1 \geq \tilde{m}+k+1$.

Case 2: $k = (N - 1)(\tilde{m} + 1)$

This means $\mathrm{rk}(\mathcal{D}_N^{(\tilde{m}+1)}(z_*)) = 0$, i.e. $\mathcal{D}_N^{(\tilde{m}+1)}(z_*) \equiv 0$. Consequently $\kappa_j^{(i)} \geq 1$ for $j = 1, \ldots, \tilde{m}+1$ and $i = 1, \ldots N$ and thus, all $N \cdot \frac{(\tilde{m}+1)(\tilde{m}+2)}{2} \geq N(\tilde{m}+2)$ polynomial entries are no fixed zeros but have the common zero z_*, which leads, according to Corollary 2, to a cardinality of $O(|X(N)| \cdot t^{N(\tilde{m}+2)-1}) = O(|X(N)| \cdot t^{k+\tilde{m}+1})$ since $N(\tilde{m}+2) - 1 = (N - 1)(\tilde{m}+2) + \tilde{m}+1 > k + \tilde{m}+1$.

Case 3: $k \leq (N - 1)(\tilde{m} + 1) - 1 \Leftrightarrow \tilde{m}(N - 1) \geq k + 2 - N$

Case 3 is divided into three subcases.

Case 3.1: $d^{(1)}_{m-\tilde{m},m-\tilde{m}}(z_*) = \cdots = d^{(N)}_{m-\tilde{m},m-\tilde{m}}(z_*) = 0$

That these polynomials have a common zero contributes a factor of $O(t^{N-1})$ to the cardinality. Additionally, these polynomials cannot be identically 1, which means $1 \leq \kappa^{(i)}_{m-(m-\tilde{m})+1} = \kappa^{(i)}_{\tilde{m}+1}$ (in particular, there is no simple form) and contributes a factor of $O(t^{N\tilde{m}})$ to the cardinality; see (2.1). In summary, the cardinality is $O(|X(N)| \cdot t^{\tilde{m}+1+k})$ since $N - 1 + N\tilde{m} \geq N - 1 + \tilde{m} + k + 2 - N = \tilde{m} + 1 + k$.

Case 3.2: $d^{(1)}_{m-\tilde{m},m-\tilde{m}}(z_*), \ldots, d^{(l-1)}_{m-\tilde{m},m-\tilde{m}}(z_*) \neq 0$ and $d^{(l)}_{m-\tilde{m},m-\tilde{m}}(z_*) = \cdots = d^{(N)}_{m-\tilde{m},m-\tilde{m}}(z_*) = 0$ for some $l \in \{2, \ldots, N\}$

All entries of row $(l - 2)(\tilde{m} + 1) + 1$ of $\mathcal{D}_N^{(\tilde{m}+1)}(z_*)$ but $d^{(l-1)}_{m-\tilde{m},m-\tilde{m}}(z_*) \neq 0$ are equal to zero. Hence, deleting the row and column of this entry decreases the rank by 1. After that, for $l \geq 3$, row $(l - 3)(\tilde{m} + 1) + 1$ of the remaining matrix consists only of zeros but $d^{(l-2)}_{m-\tilde{m},m-\tilde{m}}(z_*) \neq 0$ and the procedure could be iterated until all rows and columns of the entries $d^{(1)}_{m-\tilde{m},m-\tilde{m}}(z_*), \ldots, d^{(l-1)}_{m-\tilde{m},m-\tilde{m}}(z_*)$ are deleted and the rank is decreased by $l - 1$. Moreover, the entries of the rows $(l - 1 + j)(\tilde{m} + 1) + 1$ for $j = 0, \ldots, N - l - 1$ of $\mathcal{D}_N^{(\tilde{m}+1)}$ that are no fixed zeros are contained in the set $\{d^{(l)}_{m-\tilde{m},m-\tilde{m}}, \ldots, d^{(N)}_{m-\tilde{m},m-\tilde{m}}\}$. Thus, these rows consist only of zeros at z_* and could be deleted without changing the rank. Deleting also the columns of these entries, could only decrease the rank and one ends up with $\mathcal{D}_N^{(\tilde{m})}(z_*)$, which, consequently, has rank at most $(N-1)(\tilde{m}+1) - k - l + 1 = (N-1)\tilde{m} - (k+l-N)$. Per induction, this leads to a cardinality of $O(|X(N)| \cdot t^{\tilde{m}+k+l-N-1})$.

Since each z_* such that $\mathcal{D}_N^{(\tilde{m})}(z_*)$ is not of full row rank is a common zero of all full size subminors of $\mathcal{D}_N^{(\tilde{m})}$, it is, in particular, a zero of $\prod_{i=1}^{N} \det(D_i^{(\tilde{m})})$. Therefore, for each $\mathcal{D}_N^{(\tilde{m})}$, there exist only finitely many such $z_* \in \overline{\mathbb{F}}$. Consequently, one could regard z_* as already fixed when considering the further conditions $d_{m-\tilde{m},m-\tilde{m}}^{(l)}(z_*) = \cdots = d_{m-\tilde{m},m-\tilde{m}}^{(N)}(z_*) = 0$. Thus, these conditions contribute the factor $t^{N-l+1+\tilde{m}(N-l+1)}$ to the cardinality. Here, the summand $\tilde{m}(N-l+1)$ is due to the fact that $d_{m-\tilde{m},m-\tilde{m}}^{(i)} \neq 1$ for $i = l, \ldots, N$; see (2.1). In summary, the cardinality is $O(|X(N)| \cdot t^{\tilde{m}+k+\tilde{m}(N-l+1)}) = O(|X(N)| \cdot t^{\tilde{m}+k+1})$.

Case 3.3: $d_{\tilde{m}-m,\tilde{m}-m}^{(i)}(z_*) \neq 0$ for $i = 1, \ldots, N$

From claim 1 with $D_i^{(\tilde{m}+1)} = \begin{bmatrix} v_i & 0 \\ w_i & D_i^{(\tilde{m})} \end{bmatrix}$ and $v_i = d_{\tilde{m}-m,\tilde{m}-m}^{(i)}$ for $i = 1, \ldots, N$, one knows $\mathrm{rk}(\mathcal{D}_N^{(\tilde{m}+1)}(z_*)) = \mathrm{rk}(r(z_*) \, \mathcal{D}_N^{(\tilde{m})}(z_*)) + N - 1$, where

$$r = \begin{pmatrix} w_1 - \frac{v_1}{v_2} w_2 \\ \cdots \\ (-1)^N (\frac{w_{N-1}}{v_{N-1}} - \frac{w_N}{v_N}) v_1 \end{pmatrix} \in \mathbb{F}^{(N-1)\tilde{m}}[z].$$

It follows $\mathrm{rk}(r(z_*) \, \mathcal{D}_N^{(\tilde{m})}(z_*)) \leq (N-1)\tilde{m} - k$. If $\mathrm{rk}(\mathcal{D}_N^{(\tilde{m})}(z_*)) \leq (N-1)\tilde{m} - k - 1$, one knows per induction that the cardinality is $O(|X(N)| \cdot t^{\tilde{m}+k})$. If $\mathcal{D}_N$ is not of simple form, one has an additional factor of at most t, no matter if $\mathcal{D}_N^{(\tilde{m})}$ is of simple form or not.

If $\mathrm{rk}(\mathcal{D}_N^{(\tilde{m})}(z_*)) = (N-1)\tilde{m} - k$, one knows that the cardinality is $O(|X(N)| \cdot t^{\tilde{m}+k-1})$ and additionally, that $r(z_*)$ lies in the column span of $\mathcal{D}_N^{(\tilde{m})}(z_*)$. Hence, one has to show that this second condition leads to an additional factor for the probability that is $O(t)$. As seen above, for each $\mathcal{D}_N^{(\tilde{m})}$, there exist only finitely many $z_* \in \overline{\mathbb{F}}$ with $\mathrm{rk}(\mathcal{D}_N^{(\tilde{m})}(z_*)) = (N-1)\tilde{m} - k$. Consequently, one has to consider just the case that z_* and $\mathcal{D}_N^{(\tilde{m})}$ are fixed and the vector $r(z_*)$ lies in the column span of $\mathcal{D}_N^{(\tilde{m})}(z_*)$. If $(N-1)\tilde{m} - k < 0$, the cardinality is $O(|X(N)| \cdot t^{\tilde{m}+k+1})$, anyway (see case 1).

If $(N-1)\tilde{m} - k = 0$, one has $\mathcal{D}_N^{(\tilde{m})}(z_*) \equiv 0$ and $r(z_*) \equiv 0$. Hence, $\kappa_j^{(i)} \geq 1$ for $i = 1, \ldots, N$ and $j = 1, \ldots, \tilde{m}$, and thus, the vectors $w_1, \ldots, w_N$ contain no entries that are fixed (to zero) by degree conditions. Furthermore, one has, amongst others, $w_1(z_*) = \frac{v_1}{v_2} w_2(z_*)$, which means, in particular, that the first component of w_1 is fixed by the other polynomials, which contributes a factor that is $O(t)$ to the cardinality; see Lemma 7 (b)).

If $(N-1)\tilde{m} - k > 0$, one could choose $(N-1)\tilde{m} - k$ linearly independent rows in $\mathcal{D}_N^{(\tilde{m})}(z_*)$. If there exist $i \in \{1, \ldots, N-1\}$ and $j \in \{1, \ldots, \tilde{m}\}$ such that the j-th components of w_i and w_{i+1} are fixed to zero by degree conditions, which is the case if and only if $\kappa_{\tilde{m}+1-j}^{(i)} = \kappa_{\tilde{m}+1-j}^{(i+1)} = 0$, row $\tilde{m}(i-1) + j$ has ones in the positions $\tilde{m}(i-1) + j$

and $\tilde{m}i + j$ and zeros, elsewhere. Thus, all these rows are linearly independent and one could assume without restriction that they are contained in the choosen set of linearly independent rows. Permute the rows of $[r \; \mathcal{D}_N^{(\tilde{m})}]$ in such way that the entries of the choosen rows of $\mathcal{D}_N^{(\tilde{m})}$ are contained in the rows $1, \ldots, (N-1)\tilde{m} - k$, which we call the upper part, while the other rows of $\mathcal{D}_N^{(\tilde{m})}$ should be called the lower part. Clearly, interchanging rows does not change the rank of the whole matrix. In the following, $[r \; \mathcal{D}_N^{(\tilde{m})}]$ should denote the matrix with the already interchanged rows. Note that by this interchanging process the Hermite form is lost but this does not matter for the following considerations.

Next, delete $\tilde{m} + k$ columns of $\mathcal{D}_N^{(\tilde{m})}(z_*)$ such that the remaining entries of the upper part form an invertible matrix, denoted by $\overline{D}$. The matrix consisting of the remaining entries of the lower part should be denoted by $\underline{D}$. Analogously, denote the corresponding parts of r by $\overline{r} \in \mathbb{F}[z]^{(N-1)\tilde{m}-k}$ and $\underline{r} \in \mathbb{F}[z]^k$, respectively. Since the column rank of $\begin{pmatrix} \overline{D}(z_*) \\ \underline{D}(z_*) \end{pmatrix}$ is still $(N-1)\tilde{m} - k$, its column span is equal to the column span of $\mathcal{D}_N^{(\tilde{m})}(z_*)$ and therefore, $r(z_*)$ is contained in it. Hence, there exists $\lambda \in \overline{\mathbb{F}}^{(N-1)\tilde{m}-k}$ with $\begin{pmatrix} \overline{D}(z_*)\lambda \\ \underline{D}(z_*)\lambda \end{pmatrix} = \begin{pmatrix} \overline{r}(z_*) \\ \underline{r}(z_*) \end{pmatrix}$, i.e. $\underline{r}(z_*) = \underline{D}(z_*)\overline{D}^{-1}(z_*)\overline{r}(z_*)$. Denote the last row of $\underline{D}\overline{D}^{-1}$ by $d_1, \ldots d_{(N-1)\tilde{m}-k}$. Then $\underline{r}_k(z_*) = \sum_{l=1}^{(N-1)\tilde{m}-k} d_l \overline{r}_l(z_*)$. Moreover, $w_{i,j}$ should denote the j-th component of the vector $w_i \in \mathbb{F}[z]^{\tilde{m}}$. Thus, $\underline{r}_k = (-1)^i \left(\frac{w_{i,j}}{v_i} - \frac{w_{i+1,j}}{v_{i+1}} \right) v_1$ for some $i \in \{1, \ldots, N-1\}$ and $j \in \{1, \ldots, \tilde{m}\}$. The polynomials $w_{i,j}$ and $w_{i,j+1}$ could not both be fixed to zero due to degree conditions since otherwise $(-1)^i \left(\frac{w_{i,j}}{v_i} - \frac{w_{i+1,j}}{v_{i+1}} \right) v_1$ would belong to $\overline{r}$ per construction of upper and lower part. Assume without restriction that $w_{i,j}$ is no fixed zero.

First, consider the case that $(-1)^i \left(\frac{w_{i,j}}{v_i} - \frac{w_{i-1,j}}{v_{i-1}} \right) v_1$ is not contained in $\overline{r}$ and hence, $w_{i,j}$ is not contained in the term for any entry of $\overline{r}$. Then, one could choose all polynomial entries of $\mathcal{D}_N^{(\tilde{m}+1)}$ but $w_{i,j}$ arbitrarily, which effects that $w_{i,j}(z_*)$ is fixed. However, this contributes a factor of $O(t)$ to the cardinality; see Lemma 7 (b). If $(-1)^i \left(\frac{w_{i,j}}{v_i} - \frac{w_{i-1,j}}{v_{i-1}} \right) v_1$ is contained in $\overline{r}$, assume without restriction that it equals $\overline{r}_1$. Then, one has

$$\frac{w_{i,j}}{v_i}(1 - d_1)v_1(z_*) = \frac{w_{i+1,j}}{v_{i+1}}v_1(z_*) - \frac{w_{i-1,j}}{v_{i-1}}d_1v_1(z_*) + (-1)^i \sum_{l=2}^{(N-1)\tilde{m}-k} d_l \overline{r}_l(z_*).$$

$$(2.21)$$

Consider $d_1(z_*) = \sum_{l=1}^{(N-1)\tilde{m}-k} \underline{D}_{k,l} \overline{D}_{l,1}^{-1}(z_*)$.

Case 3.3.1: The entries of κ are so that $d_1 \equiv 0$ (by degree conditions).

Here, one has, in particular, $d_1(z_*) \neq 1$ and could, therefore, solve equation (2.21) with respect to $w_{i,j}(z_*)$. Hence, one has a factor that is $O(t)$ for the cardinality and is done.

Case 3.3.2: The entries of κ are not so that they imply $d_1 \equiv 0$.

If $d_1(z_*) = 0$, which also implies that one could solve equation (2.21) with respect to $w_{i,j}(z_*)$ and consequently, is done as well, there exists l_* such that neither $\underline{D}_{k,l_*} \equiv 0$ nor $\overline{D}_{l_*,1}^{-1} \equiv 0$ due to degree restrictions (caused by the values of κ). Thus, either $\overline{D}_{l_*,1}^{-1}(z_*) = 0$, which leads to a factor which is $O(t)$ for the cardinality, or one could solve the equation $d_1(z_*) = 0$ with respect to $\underline{D}_{k,l_*}(z_*)$ (that is no fixed 1 per construction of upper and lower part), which provides the factor $O(t)$, too. Therefore, the probability that $d_1(z_*) = 0$ if not $d_1 \equiv 0$ due to degree conditions, is $O(t)$. Hence, it only remains to investigate what happens if $d_1(z_*) \neq 0$, which is true with a probability of $1 - O(t)$ in the considered case. This implies that the probability that $\mathrm{rk}(\mathcal{D}_N^{(\tilde{m})}(z_*)) = (N-1)\tilde{m} - k$ under the condition $d_1(z_*) \neq 0$ is $\frac{O(t^{\tilde{m}+k-1})}{1-O(t)} = O(t^{\tilde{m}+k-1})$.

Per construction of $\overline{D}$ and $\underline{D}$, it does not influence the condition $\mathrm{rk}(\mathcal{D}_N^{(\tilde{m})}(z_*)) = (N-1)\tilde{m} - k$, which nonzero value is taken by $d_1(z_*)$. This is true since $\overline{D}(z_*)$ is invertible and therefore, the rows of $\underline{D}(z_*)$ are linearly dependent on the rows of $\overline{D}(z_*)$, anyway. Moreover, multiplying a row by a nonzero factor, does not influence linear dependence, i.e. does not influence the number of possibilities for the entries of $\mathcal{D}_N^{(\tilde{m})}$ which are not contained in $\overline{D}$ or $\underline{D}$. If $d_1(z_*) \neq 1$, one could solve equation (2.21) with respect to $w_{i,j}(z_*)$ and is done.

If $1 = d_1(z_*) = \sum_{l=1}^{(N-1)\tilde{m}-k} \underline{D}_{k,l} \overline{D}_{l,1}^{-1}(z_*)$, there exists $l_0 \in \{1, \ldots, (N-1)\tilde{m}-k\}$ such that $\overline{D}_{l_0,1}^{-1}(z_*) \neq 0$ and $\underline{D}_{k,l_0}$ is no fixed zero (it cannot be a fixed 1 per construction of upper and lower part). Consequently, one could solve the above equation with respect to $\underline{D}_{k,l_0}(z_*)$. Because it follows from the preceding considerations that the condition $d_1(z_*) = 1$ is independent from the condition $\mathrm{rk}(\mathcal{D}_N^{(\tilde{m})}(z_*)) = (N-1)\tilde{m}-k$, one gets an additional factor that is $O(t)$ for the cardinality. As in previous cases, the cardinality is decreased by a factor of at most t if one has no simple form and thus, all cases are finished.

Note that it is sufficient to consider these three cases since the order of $D_1, \ldots, D_N$ is not relevant for the property to be mutually left coprime. Therefore, the proof of claim 2 is complete.

Using claim 2 with $\tilde{m} = m$ and $k = 1$, completes the first part of this proof.

Next, one needs to compute the probability for the case that $\mathcal{D}_N \subset A_{N+1}(N)$ is of simple form, i.e. the case that $D_i = \begin{bmatrix} I_{m-1} & 0 \\ d_1^{(i)} \cdots d_{m-1}^{(i)} & d_m^{(i)} \end{bmatrix}$ for $i = 1, \ldots, N$.

Claim 3:

For $\mathcal{D} :=$

$$
\begin{bmatrix}
d_1^{(1)} - d_1^{(2)} & \cdots & d_{m-1}^{(1)} - d_{m-1}^{(2)} & d_m^{(1)} & d_m^{(2)} & & \\
d_1^{(3)} - d_1^{(2)} & \cdots & d_{m-1}^{(3)} - d_{m-1}^{(2)} & & d_m^{(2)} & d_m^{(3)} & \\
\vdots & & \vdots & & & \ddots & \ddots \\
(-1)^N(d_1^{(N-1)} - d_1^{(N)}) & \cdots & (-1)^N(d_{m-1}^{(N-1)} - d_{m-1}^{(N)}) & & & d_m^{(N-1)} & d_m^{(N-1)}
\end{bmatrix}
$$

it holds $\mathrm{rk}(\mathcal{D}_N) = \mathrm{rk}(\mathcal{D}) + (m-1)(N-1)$.

Proof of claim 3:
One proceeds as in the proof of claim 1 (with $v_i = 1$) and achieves:

$$\text{rk}(\mathcal{D}_N) =$$

$$\text{rk}\begin{bmatrix} w_1 - w_2 & D_1^{(m-1)} & D_2^{(m-1)} & & \\ w_3 - w_2 & & D_2^{(m-1)} & D_3^{(m-1)} & \\ \vdots & & & \ddots & \ddots \\ (-1)^N(w_{N-1} - w_N) & & & D_{N-1}^{(m-1)} & D_N^{(m-1)} \end{bmatrix} + N - 1,$$

where $w_i = (0, \ldots, 0, d_1^{(i)})^\top \in \mathbb{F}[z]^{m-1}$ and $D_i^{(m-1)} \in \mathbb{F}[z]^{(m-1) \times (m-1)}$ is in simple form for $i = 1, \ldots, N$. One iterates this procedure $m - 1$ times and since one always adds the first row of a block to rows further down, the first column of the whole matrix is not affected. Deleting the corresponding row only deletes one zero in each of the vectors w_i. After $m - 1$ iterations, one ends up with the statement of claim 3 and thus, claim 3 is proven.
Consequently, for simple form, $D_1, \ldots, D_N$ are not mutually left coprime if and only if there exist $z_* \in \overline{\mathbb{F}}$ and $\xi \in \overline{\mathbb{F}}^{1 \times (N-1)} \setminus \{0\}$ such that $\xi \mathcal{D}(z_*) = 0$, which is equivalent to

$$\xi_1 d_i^{(1)}(z_*) - (\xi_1 + \xi_2) d_i^{(2)}(z_*) + \cdots + (-1)^N(\xi_{N-2} + \xi_{N-1}) d_i^{(N-1)}(z_*)+$$
$$+(-1)^{N+1} \xi_{N-1} d_i^{(N)}(z_*) = 0 \quad \text{for } i = 1, \ldots m - 1$$
$$\xi_1 d_m^{(1)}(z_*) = 0$$
$$(\xi_{i-1} + \xi_i) d_m^{(i)}(z_*) = 0 \quad \text{for } i = 2, \ldots, N - 1$$
$$\xi_{N-1} d_m^{(N)}(z_*) = 0. \tag{2.22}$$

Next, define $\tilde{A}_{N+1}(N)$ as the subset of $X(N)$ for which there exists $z_* \in \overline{\mathbb{F}}$ such that $\mathcal{D}_N(z_*)$ is singular and $\det(D_i(z_*)) = 0$ for $i = 1, \ldots, N$. From Remark 11 it follows $A_{N+1}(N) \subset \tilde{A}_{N+1}(N)$. In the following, we compute the probability of $\tilde{A}_{N+1}(N)$ for simple form, i.e. the probability that there exist $z_* \in \overline{\mathbb{F}}$ and $\xi \in \mathbb{F}(z_*)^{1 \times (N-1)} \setminus \{0\}$ with $d_m^{(i)}(z_*) = 0$ for $i = 1, \ldots, N$, such that the first $m - 1$ equations of (2.22) are fulfilled. Firstly, that these N polynomials have a common zero gives a factor to the cardinality of $O(t^{N-1})$.
If $m < N - 1$, this leads to a cardinality that is $O(|X(N)| \cdot t^{m+1})$.
To consider the case $m \geq N-1$, one sorts the possible values for z_* with respect to the degree of their minimal polynomial und sets $g := g_{z_*}$. Then, $\xi \in (\mathbb{F}^g)^{1 \times (N-1)} \setminus \{0\}$, where $\mathbb{F}^g$ denotes the extension field with t^{-g} elements. Hence, there are $t^{-g(N-1)} - 1$ possibilities for the choice of ξ. Since there exists $i \in \{1, \ldots, N - 1\}$ with $\xi_i \neq 0$, at least one element of $\{\xi_1, \xi_1 + \xi_2, \ldots, \xi_{N-2} + \xi_{N-1}, \xi_{N-1}\}$ is unequal to zero and thus, there exists $j_0 \in \{1, \ldots, N\}$ such that one could solve equations 1 to $m - 1$ of (2.22) with respect to $d_i^{(j_0)}(z_*)$ for $i = 1, \ldots, m - 1$. Assume without restriction that $j_0 = 1$. If the other entries of $\mathcal{D}_N$ as well as z_* are fixed, for $\xi, \hat{\xi} \in (\mathbb{F}^g)^{1 \times (N-1)} \setminus \{0\}$, this

leads to the same values for $\xi_1 d_i^{(1)}(z_*)$ if and only if

$$(\xi_1 + \xi_2)d_i^{(2)}(z_*) + \cdots + (-1)^{N+1}\xi_{N-1}d_i^{(N)}(z_*) =$$
$$(\hat{\xi}_1 + \hat{\xi}_2)d_i^{(2)}(z_*) + \cdots + (-1)^{N+1}\hat{\xi}_{N-1}d_i^{(N)}(z_*)$$

for $i = 1, \ldots, m - 1$. But these equations hold if and only if the vector

$$\begin{pmatrix} \xi_1 + \xi_2 \\ \vdots \\ (-1)^{N-1}\xi_{N-1} \end{pmatrix} - \begin{pmatrix} \hat{\xi}_1 + \hat{\xi}_2 \\ \vdots \\ (-1)^{N-1}\hat{\xi}_{N-1} \end{pmatrix} \quad \text{is contained in the kernel of}$$

$$D(z_*) := \begin{bmatrix} d_1^{(2)} & \cdots & d_1^{(N)} \\ \vdots & & \vdots \\ d_{m-1}^{(2)} & \cdots & d_{m-1}^{(N)} \end{bmatrix} (z_*).$$

The probability that the column rank of $D(z_*)$ is $\min(N - 1, m - 1)$ is equal to $1 - O(t)$. This is true because the probability that a full size minor of this matrix is zero at a fixed value z_* is equal to $O(t)$ if one chooses $d_j^{(i)}$ with $\deg(d_j^{(i)}) < n_i$ for $j = 1, \ldots m - 1$ and $i = 1, \ldots, N$ randomly; this follows from Lemma 7 (a) and (b) because that a minor is zero implies that either one of the involved polynomial entries is zero or one of the entries is fixed by the others. Consequently, for $m \geq N$, i.e. $\min(N - 1, m - 1) = N - 1$, the probability that the kernel of $D(z_*)$ is zero is equal to $1 - O(t)$. Therefore, the probability that $\xi = \hat{\xi}$ is equal to $1 - O(t)$, too. For $m = N - 1$, one has $\min(N - 1, m - 1) = N - 2$ and thus, the probability that the kernel has dimension one is equal to $1 - O(t)$. This means that only ξ that differ by a nonzero scalar factor lead to the same solution. But if one multiplies ξ by a factor from $\mathbb{F}^g \setminus \{0\}$, the set of possible values for the D_i which fulfill (2.22) does not change, anyway.

In summary, with probability $1 - O(t)$, one has $\frac{t^{-g(N-1)}-1}{t^{-g}-1} = \sum_{k=0}^{N-2}(t^{-g})^k =$ $= t^{-g(N-2)}(1 - O(t))$ possibilities for ξ and according to Lemma 7 (a) and (b), for each ξ, there are $O(|X(N)| \cdot t^{g(N-1)+m-1})$ possibilities for $\mathcal{D}_N$. Hence, the probability is $O(t^{g+m-1}) = O(t^{m+1})$ for $g \geq 2$. Since one already knows that f_{z_*} divides $d_m^{(i)}$ for $i = 1, \ldots, N$, one has $g \leq \min(n_1, \ldots, n_N)$, i.e. there are only finitely many possibilities for g. Consequently, only the case $g = 1$ is relevant for the computation of the coefficient of t^m. Here, one has $t^{-(N-2)}(1 - O(t))$ possibilities for ξ and according to Lemma 7 (a) and (b), t^{N+m-2} possibilities for z_* and $\mathcal{D}_N$. Hence, the probability of $\tilde{A}_{N+1}(N)$ is $t^m + O(t^{m+1})$.

Next, we show that for simple form, it holds $|A_{N+1}(N)| = |\tilde{A}_{N+1}(N)| + O(|X(N)| \cdot t^{m+1})$, which imples $|A_{N+1}(N)| = |X(N)| \cdot O(t^{m+1})$ if $m < N - 1$ and $|A_{N+1}(N)| = |X(N)| \cdot (t^m + O(t^{m+1}))$ if $m \geq N - 1$, i.e.

$$|A_{N+1}(N)| = |X(N)| \cdot \left(\sum_{i=N-1}^{\min(m, N-1)} t^m + O(t^{m+1}) \right).$$

We prove this by showing that for simple form, $M^C(N) := X(N) \setminus mut(N)$ and $A := (M^C(N) \setminus A_{N+1}(N)) \cap \tilde{A}_{N+1}(N)$, one has $|A| = O(|X(N)| \cdot t^{m+1})$. It holds that $(D_1, \ldots, D_N) \in A$ if and only if there exist $z_*, \tilde{z}_* \in \mathbb{F}$ such that $d_m^{(i)}(z_*) = 0$ for $i = 1, \ldots, N$ and the first $m - 1$ equations of (2.22) are fulfilled for z_* and there exists a subset of $N - 1$ matrices which fulfil equations (2.22) at $\tilde{z}_*$. Since the number of choices for this subset is equal to N and therefore finite, it follows from preceding computations that the probability of $M^C(N) \setminus A_{N+1}(N)$ (i.e. of the condition concerning $\tilde{z}_*$) is $O(t^m)$. Without restriction, let the mentioned subset be $\{D_1, \ldots, D_{N-1}\}$. If $z_* = \tilde{z}_*$, one has, amongst others, the additional condition $d_m^{(N)}(\tilde{z}_*) = 0$, which gives a factor that is $O(t)$ for the probability, according to Lemma 7 (a). If $z_* \neq \tilde{z}_*$, one has the additional conditions that $d_m^{(i)}(z_*) = 0$ for $i = 1, \ldots, N$, which contributes a factor that is $O(t^{N-1}) = O(t)$. Consequently, in summary, one has that the probability of A is $O(t^{m+1})$, which is what we wanted to show.

It remains to compute $|A_I(N)|$. From claim 2, one already knows $|A_I(N)| = |X(N)| \cdot O(t^m)$. First consider $A_i(N) \cap A_j(N)$, i.e. $I = \{i, j\}$ with $i \neq j$, and assume without restriction $i = 1$ and $j = N$. It holds $(D_1, \ldots, D_N) \in A_1(N) \cap A_N(N)$ if and only if $\{D_2, \ldots, D_N\}$ and $\{D_1, \ldots, D_{N-1}\}$ are not mutually left coprime. Since the condition that $\{D_2, \ldots, D_N\}$ are not mutually left coprime causes already a factor for the probability that is $O(t^{m+1})$ if $\mathcal{D}_N$ is not of simple form, it is only necessary to consider simple form. Denote by $\hat{A}_{1,N}(N)$ the subset of $X(N)$ for which $\{D_2, \ldots, D_{N-1}\}$ are not mutually left coprime and write $|A_1(N) \cap A_N(N)| = |A_1(N) \cap A_N(N) \cap \hat{A}_{1,N}(N)| + |A_1(N) \cap A_N(N) \cap \hat{A}_{1,N}^C(N)|$, where $\hat{A}_{1,N}^C(N)$ denotes the complementary set $X(N) \setminus \hat{A}_{1,N}(N)$. Moreover, denote by $\mathcal{D}_N^{(1)}$ the matrix that is achieved if the first m rows and columns of $\mathcal{D}_N$ are deleted. Analogously, denote by $\mathcal{D}_N^{(N)}$ the matrix that is achieved if the last m rows and columns of $\mathcal{D}_N$ are deleted. If $\mathcal{D}_N \in A_1(N) \cap A_N(N)$ is of simple form, one knows that equations (2.22) are valid for $\mathcal{D}_N^{(1)}$ as well as for $\mathcal{D}_N^{(N)}$. Denote the corresponding ξ, z_* and g_{z_*} by $\xi^{(1)}, z_*^{(1)}, g^{(1)}$ and $\xi^{(N)}, z_*^{(N)}, g^{(N)}$, respectively. If $\mathcal{D}_N \in A_1(N) \cap A_N(N) \cap \hat{A}_{1,N}^C(N)$, one has $\xi_{N-2}^{(1)} \neq 0$ as well as $\xi_1^{(N)} \neq 0$ and therefore, $d_m^{(N)}(z_*^{(1)}) = d_m^{(1)}(z_*^{(N)}) = 0$ (see proof of Remark 11). For fixed $z_*^{(1)}$ and $z_*^{(N)}$ (where without restriction $\deg(d_m^{(N)}) = n_m \geq g^{(1)}$ and $\deg(d_m^{(1)}) = n_1 \geq g^{(N)}$ since otherwise, $A_1(N) \cap A_N(N) \cap \hat{A}_{1,N}^C(N) = \emptyset$, anyway), this contributes a factor of $t^{g^{(1)}}$ for the probability that $D_2, \ldots, D_N$ are not mutually left coprime and a factor of $t^{g^{(N)}}$ for the probability that $D_1, \ldots, D_{N-1}$ are not mutually left coprime, respectively. Thus, the other equations of (2.22) for $\mathcal{D}_N^{(1)}$ contribute a factor that is $O(t^{m-g^{(1)}})$ and the other equations of (2.22) for $\mathcal{D}_N^{(N)}$ contribute a factor that is $O(t^{m-g^{(N)}})$. Assume without restriction $g^{(1)} \geq g^{(N)}$. Then, one has a contribution to the probability that is $O(t^{m-g^{(1)}})$ by the equations for $\mathcal{D}_N^{(1)}$ and $\mathcal{D}_N^{(N)}$ but $d_m^{(N)}(z_*^{(1)}) = d_m^{(1)}(z_*^{(N)}) = 0$ and the additional factor $t^{g^{(1)}+g^{(N)}}$ for these equations. Hence, in summary,
$$|A_1(N) \cap A_N(N) \cap \hat{A}_{1,N}^C(N)| = O(|X(N)| \cdot t^{m+g^{(N)}}) = O(|X(N)| \cdot t^{m+1}).$$

Consequently, $|A_1(N) \cap A_N(N)| = |A_1(N) \cap A_N(N) \cap \hat{A}_{1,N}(N)| + O(|X(N)| \cdot t^{m+1}) = |\hat{A}_{1,N}(N)| + O(|X(N)| \cdot t^{m+1})$. Therefore,

$$|A_i(N) \cap A_j(N)| = |X(N)| \cdot (1 - P_m(N-2) + O(t^{m+1}))$$

for $i, j \in \{1, \ldots, N\}$ with $i \neq j$.

Next, it is shown per induction with respect to $|I|$ that $|A_I(N)| = |\hat{A}_I(N)| + O(|X(N)| \cdot t^{m+1})$ for $2 \leq |I| \leq N - 2$, where $\hat{A}_I(N)$ denotes the subset of $X(N)$ for which the $N - |I|$ matrices from the set $\{D_i\}_{i \in \{1,\ldots,N\} \setminus I}$ are not mutually left coprime. The proof of the corresponding base clause has already been done in the preceding paragraph. For $|I| = k$ with $3 \leq k \leq N - 2$, assume without restriction that $I = \{N - k + 1, \ldots, N\}$. Since $\hat{A}_I(N) \subset A_I(N)$, one has

$$A_I(N) = A_{N-k+2,\ldots,N}(N) \cap A_{N-k+1,N-k+3,\ldots,N}(N) =$$
$$= \hat{A}_{N-k+2,\ldots,N}(N) \cap \hat{A}_{N-k+1,N-k+3,\ldots,N}(N) +$$
$$+ (A_{N-k+2,\ldots,N}(N) \cap A_{N-k+1,N-k+3,\ldots,N}(N)) \setminus$$
$$(\hat{A}_{N-k+2,\ldots,N}(N) \cap \hat{A}_{N-k+1,N-k+3,\ldots,N}(N)).$$

Furthermore,

$$|(A_{N-k+2,\ldots,N}(N) \cap A_{N-k+1,N-k+3,\ldots,N}(N)) \setminus$$
$$(\hat{A}_{N-k+2,\ldots,N}(N) \cap \hat{A}_{N-k+1,N-k+3,\ldots,N}(N))| \leq$$
$$\leq |(A_{N-k+2,\ldots,N}(N) \cap A_{N-k+1,N-k+3,\ldots,N}(N)) \setminus \hat{A}_{N-k+2,\ldots,N}(N)| +$$
$$+ |(A_{N-k+2,\ldots,N}(N) \cap A_{N-k+1,N-k+3,\ldots,N}(N)) \setminus \hat{A}_{N-k+1,N-k+3,\ldots,N}(N)| \leq$$
$$\leq |A_{N-k+2,\ldots,N}(N) \setminus \hat{A}_{N-k+2,\ldots,N}(N)| +$$
$$+ |A_{N-k+1,N-k+3,\ldots,N}(N) \setminus \hat{A}_{N-k+1,N-k+3,\ldots,N}(N)| = O(|X(N)| \cdot t^{m+1})$$

per induction since $|\{N - k + 2, \ldots, N\}| = |\{N - k + 1, N - k + 3, \ldots, N\}| = k - 1$. Moreover, it holds $\hat{A}_{N-k+2,\ldots,N}(N) = \hat{A}_{N-k+2}(N-k+2) = A_{N-k+2}(N-k+2)$ and $\hat{A}_{N-k+1,N-k+3,\ldots,N}(N) = \hat{A}_{N-k+1}(N-k+2) = A_{N-k+1}(N-k+2)$. Consequently,

$$|A_I(N)| = |A_{N-k+2}(N-k+2) \cap A_{N-k+1}(N-k+2)| + O(|X(N)| \cdot t^{m+1}) =$$
$$= |A_{N-k+1,N-k+2}(N-k+2)| + O(|X(N)| \cdot t^{m+1}) =$$
$$= |\hat{A}_{N-k+1,N-k+2}(N-k+2)| + O(|X(N)| \cdot t^{m+1}) =$$
$$= |\hat{A}_I(N)| + O(|X(N)| \cdot t^{m+1}). \tag{2.23}$$

Here, the third equation follows from the base clause.

For $|I| = N - 1$, assume without loss of generality that $I = \{2, \ldots, N\}$. Analogous to the first line of (2.23) (with $k = N - 1$), one gets $|A_I(N)| = |S_{12} \cap S_{13}| + O(|X(N)| \cdot t^{m+1})$, where S_{12} and S_{13} are the subsets of $X(N)$ for which D_1, D_2 and D_1, D_3 are not left coprime, respectively. In the proof of Theorem 20, it has been shown that $|S_{12} \cap S_{13}| = O(|X(N)| \cdot t^{m+1})$. Therefore, $|A_I| = O(|X(N)| \cdot t^{m+1})$ for $|I| = N - 1$

and consequently, $|A_I| = O(|X(N)| \cdot t^{m+1})$ for $|I| = N$, too.

In summary, one has $|A_I| = |X(N)| \cdot (1 - P_m(N - |I|) + O(t^{m+1}))$ for $|I| \leq N - 2$ and $|A_I| = |X(N)| \cdot O(t^{m+1})$ for $|I| \in \{N - 1, N\}$.

Inserting all results into (2.20), using that there are $\binom{N}{|I|}$ subset of $\{1, \ldots, N\}$ with cardinality $|I|$ and dividing by $|X(N)|$ completes the proof of the whole theorem. $\square$

The preceding theorem makes it possible to compute $P_m(N)$, which is firstly done for $N = 3, 4, 5$ in the following example.

Example 9.

For $N = 2, 3, 4$, one has the following probabilities for mutual left coprimeness:

$$P_m(3) = 1 - (3 + \sum_{i=2}^{\min(m,2)} 1)t^m + O(t^{m+1}),$$

$$P_m(4) = 1 - (6 + 4 \sum_{i=2}^{\min(m,2)} 1 + \sum_{i=3}^{\min(m,3)} 1)t^m + O(t^{m+1}),$$

$$P_m(5) = 1 - (10 + 10 \sum_{i=2}^{\min(m,2)} 1 + 5 \sum_{i=3}^{\min(m,3)} 1 + \sum_{i=4}^{\min(m,4)} 1)t^m + O(t^{m+1}).$$

Next, an explicit formula for the coefficient of t^m in $P_m(N)$ should be developed. Therefore, we write

$$P_m(N) = 1 + \left(\sum_{y \geq 1} c_y(N) \sum_{i=y}^{\min(m,y)} 1 \right) t^m + O(t^{m+1})$$

with coefficients $c_y(N) \in \mathbb{N}$, which remain to be computed.

Note that $\sum_{i=1}^{\min(m,1)} 1 = 1$ for all $m \in \mathbb{N}$. Moreover, one will see in the following that the sum over j is finite, i.e. there exists $y_0 \in \mathbb{N}$ with $c_y(N) = 0$ for $y \geq y_0$. We proceed by computing a recursion formula for $c_y(N)$ and solving it to derive an explicit formula for these coefficients. It could be seen easiliy that $c_1(N) = -\binom{N}{2}$: setting $m = 1$, leads to $P_1(N) = 1 + c_1(N)t + O(t^2)$ since $\sum_{i=y}^{\min(1,y)} 1 = 1$ for $y = 1$ and $\sum_{i=y}^{\min(1,y)} 1 = 0$ for $y > 1$. But for $m = 1$, mutual left coprimeness is equivalent to pairwise coprimeness. Hence, the statement follows from Theorem 20.

Lemma 10.

Let $c_y(N)$ be the coefficient of $\sum_{i=y}^{\min(m,y)} t^m$ in $P_m(N)$. Then it holds:

$$c_y(N) = \sum_{k=1}^{N-y-1} (-1)^{k+1} \binom{N}{k} c_y(N - k) \qquad \text{for } 1 \leq y \leq N - 2$$

$$c_{N-1}(N) = -1$$

$$c_y(N) = 0 \qquad \text{for } y \geq N.$$

Proof.
From the recursion formula for $P_m(N)$ follows

$$c_y(N) = \sum_{k=1}^{N-2} (-1)^{k+1} \binom{N}{k} c_y(N-k) - 1 \qquad \text{for } y = N-1$$

$$c_y(N) = \sum_{k=1}^{N-2} (-1)^{k+1} \binom{N}{k} c_y(N-k) \qquad \text{otherwise.}$$

Using these formulas, the lemma is shown per induction with respect to N. It is already known that $c_1(2) = -1$ and $c_y(2) = 0$ for $y \geq 2$. Hence, per induction, one knows $c_y(N-k) = 0$ for $y \geq N-k$, i.e. $k \geq N-y$. Consequently, one can choose $N-y-1$ as upper limit for the sum and the lemma is proven for $y \leq N-2$. Furthermore, for $y \geq N-1$, the sum vanishes and it remains -1 if $y = N-1$ and 0 if $y \geq N$. $\square$

Now, we are ready to solve the recursion formula of Theorem 21 and achieve an explicit formula for the probability of mutual left coprimeness.

Theorem 22.
For $m, N \geq 2$, it holds:

$$P_m(N) = 1 - \sum_{y=2}^{m+1} \binom{N}{y} t^m + O(t^{m+1}).$$

Proof.
At first, it is shown that $c_y(N) = -\binom{N}{N-y-1}$. This is done per induction with respect to N. It holds $c_{N-1}(N) = -1 = -\binom{N}{0}$ and $c_y(N) = 0 = -\binom{N}{N-y-1}$ for $y \geq N$. Moreover, these identities are sufficient to prove the claim for $N = 2$. Per induction one obtains for $1 \leq y \leq N-2$,

$$c_y(N) = \sum_{k=1}^{N-y-1} (-1)^{k+1} \binom{N}{k} c_y(N-k) = \sum_{k=1}^{N-y-1} (-1)^k \binom{N}{k} \binom{N-k}{N-k-y-1} =$$

$$= \sum_{k=1}^{N-y-1} (-1)^k \frac{N!}{k!(N-k-y-1)!(y+1)!} =$$

$$= \frac{N!}{(y+1)!} \sum_{k=1}^{N-y-1} (-1)^k \frac{1}{k!(N-k-y-1)!}. \tag{2.24}$$

For the computation of the last sum substitute $M = N-y-1$ and achieve:

$$\sum_{k=1}^{M} (-1)^k \frac{1}{k!(M-k)!} = \sum_{k=0}^{M} (-1)^k \frac{1}{k!(M-k)!} - \frac{1}{M!} =$$

$$= \frac{1}{M!} \left(\sum_{k=0}^{M} (-1)^k \binom{M}{k} - 1 \right) = -\frac{1}{M!}$$

since applying the binomial theorem shows $\sum_{k=0}^{M}(-1)^k\binom{M}{k} = 0$. Resubstitution and inserting into (2.24) leads to

$$P_m(N) = 1 - \left(\sum_{y=1}^{N-1}\binom{N}{N-y-1}\sum_{i=y}^{\min(m,y)}1\right)t^m + O(t^{m+1}).$$

The identities

$$\sum_{y=1}^{N-1}\binom{N}{y+1}\sum_{i=y}^{\min(m,y)}1 = \sum_{y=1}^{\min(N-1,m)}\binom{N}{y+1} = \sum_{y=2}^{\min(N,m+1)}\binom{N}{y} = \sum_{y=2}^{m+1}\binom{N}{y}$$

complete the proof. $\qquad\square$

Remark 13.
One can prove the preceding theorem using only the recursion formula for $P_m(N)$ and without considering the coefficients $c_y(N)$. From the recursion formula of $P_m(N)$, one can deduce a recursion formula for the coefficient $C(N)$ of t^m in $P_m(N)$ as well, which has the following form:

$$C(N) = \sum_{k=1}^{N-2}(-1)^{k+1}\binom{N}{k}C(N-k) - \sum_{i=N-1}^{\min(m,N-1)}1.$$

With the help of this formula, one can prove $C(N) = -\sum_{y=2}^{m+1}\binom{N}{y}$ per induction with respect to N. This way seems to be more straightforward than considering $c_y(N)$ but it is harder to see the solution of the recursion formula for $C(N)$ than to see the solution of the recursion formula for $c_y(N)$. Moreover, the proof per induction is harder, too.

Proof.
For $N = 2$, one has $-\sum_{y=2}^{m+1}\binom{2}{y} = -1$, which coincides with the result of Theorem 5. Moreover, per induction, one knows

$$C(N) = \sum_{k=1}^{N-2}(-1)^k\binom{N}{k}\sum_{y=2}^{m+1}\binom{N-k}{y} - \sum_{i=N-1}^{\min(m,N-1)}1 =$$

$$= \sum_{k=1}^{N-2}\sum_{y=2}^{\min(m+1,N-k)}(-1)^k\frac{N!}{k!\cdot y!\cdot(N-k-y)!} - \sum_{i=N-1}^{\min(m,N-1)}1 =$$

$$= \sum_{y=2}^{\min(m+1,N-1)}\sum_{k=1}^{N-y}(-1)^k\frac{N!}{k!\cdot y!\cdot(N-k-y)!} - \sum_{i=N-1}^{\min(m,N-1)}1 =$$

$$= -\sum_{y=2}^{\min(m+1,N-1)}\binom{N}{y} - \sum_{i=N-1}^{\min(m,N-1)}1.$$

The computation of the last step was done analogous to the transformation of (2.24). If $m \leq N - 2$, the second sum vanishes and $\min(m + 1, N - 1) = m + 1$. Thus, $C(N) = -\sum_{y=2}^{m+1} \binom{N}{y}$. If $m \geq N - 1$, the second sum is equal to 1 and $\min(m + 1, N - 1) = N - 1$. Hence, one obtains $C(N) = -\sum_{y=2}^{N-1} \binom{N}{y} - 1 = -\sum_{y=2}^{N} \binom{N}{y} = -\sum_{y=2}^{m+1} \binom{N}{y}$ since $\binom{N}{y} = 0$ for $y > N$. $\qquad\square$

Again, we want to compare the preceding result with the formula one gets for the natural density of mutual left coprimeness, which should be computed in the following. It will turn out that, as in the previous section, the problem of computing the natural density could be reduced to the calculation of the (uniform) probability that N constant matrices are mutually left coprime. Therefore, we start with the following definition:

Definition 13.

For $j \in \mathbb{N}$, denote by $W_j(N)$ the probability that $\mathcal{K}_N :=$

$$\begin{bmatrix} K_1 & K_2 & 0 & 0 \\ 0 & \ddots & \ddots & 0 \\ 0 & 0 & K_{N-1} & K_N \end{bmatrix}$$

with $K_i \in (\mathbb{F}^j)^{m \times m}$ for $i = 1, \ldots, N$ is of full row rank.

Theorem 23.
The natural density of N matrices $D_i \in \mathbb{F}[z]^{m \times m}$ for $i = 1, \ldots, N$ to be mutually left coprime is equal to $\prod_{j=1}^{\infty} W_j(N)^{\varphi_j}$.

Proof.
Similar to the proof of Theorem 17, one obtains that the N matrices are mutually left coprime if and only if $\mathcal{D}_N/(f) := \mathcal{K}_N \in (\mathbb{F}^{\deg(f)})^{(N-1)m \times Nm}$ has full row rank for every (monic) irreducible polynomial $f \in \mathbb{F}[z]$. Denote the probability for this fact by W_f. As in the proof of Theorem 17 and with the notation from there, one gets $\lim_{n \to \infty} \frac{|E_P \cap M_n|}{|M_n|} = \prod_{f \in P} W_f$. According to that proof, one needs $W_f = 1 + O(t^{2 \deg(f)})$ to conclude $\lim_{n \to \infty} \frac{|E \cap M_n|}{|M_n|} = \prod_{j=1}^{\infty} W_j(N)^{\varphi_j}$. To this end, one shows that at least 2 of the matrices K_i have zero determinant if these matrices are not mutually left coprime. If $N = 2$, this clearly is true because the matrices are left coprime if not both of them have zero determinant. For $N \geq 3$, assume without restriction that $\det(K_N) \neq 0$ (otherwise permutate the matrices K_i). Consequently, the columns of this matrix form a basis of $(\mathbb{F}^{\deg(f)})^m$ and adding appropriate linear combinations of the last m columns to the m preceding columns of $\mathcal{K}_N$ brings this

matrix to the form $\begin{bmatrix} K_1 & K_2 & 0 & \cdots & 0 \\ 0 & \ddots & \ddots & \ddots & 0 \\ 0 & 0 & K_{N-2} & K_{N-1} & 0 \\ 0 & 0 & 0 & 0 & K_N \end{bmatrix}$, which is not left prime if

and only if the submatrix consisting of the first $(N - 2)m$ rows is not left prime. Per induction, it follows that at least two of the matrices $K_1, \ldots, K_{N-1}$ have zero determinant, which gives us the desired result. Furthermore, the probability that the determinant of K_i is equal to zero is $1 - t^{m^2 \deg(f)} |GL_m(\mathbb{F}^{\deg(f)})| = O(t^{\deg(f)})$. Thus, the proof is complete. $\qquad\square$

It remains to compute $W_j(N)$. To this end, we will firstly prove a recursion formula for it.

Lemma 11.
Let $\hat{A}$ be the set of matrices K_i for which $\mathcal{K}_N$ has full row rank and $\det(K_i) = 0$ for $i = 1, \ldots, N$. Moreover, denote by $\hat{W}_j(N)$ the probability of $\hat{A}$. With $W_j(0) = W_j(1) = 1$, it holds for $N \geq 2$:

$$W_j(N) = \sum_{i=1}^{N} (-1)^{i-1} \binom{N}{i} \left(t^{jm^2} |GL_m(\mathbb{F}^j)| \right)^i W_j(N-i) + \hat{W}_j(N).$$

Proof.
If $\det(K_i) \neq 0$ for some $i \in \{1, \ldots, N\}$, one sees as in the preceding proof that $\mathcal{K}_N$ has full row rank if and only if the matrix $\mathcal{K}_{N-1}^{(i)}$ formed by the matrices from the set $\{K_1, \ldots, K_N\} \setminus \{K_i\}$ has full row rank. Using the inclusion-exclusion-principle with $A_i := \{\det(K_i) \neq 0$ and $\mathcal{K}_{N-1}^{(i)}$ is of full row rank$\}$ and $\hat{A}$, where $\hat{A} \cap A_i = \emptyset$, $P_i := \Pr(A_i)$ for $i = 1, \ldots, N$ and $P_I := \Pr\left(\bigcap_{i \in I} A_i(N) \right)$, one gets

$$W_j(N) = \sum_{I \subset \{1, \ldots, N\}} (-1)^{|I|-1} P_I + \hat{W}_j(N).$$

With the same arguments as in the preceding proof, one obtains
$\bigcap_{i \in I} A_i(N) = \{\det(K_i) \neq 0$ for $i \in I$ and $\mathcal{K}_{N-|I|}^{(I)}$ has full row rank$\}$ and therefore, $P_I = \left(t^{jm^2} |GL_m(\mathbb{F}^j)| \right)^i W_j(N-i)$ for every I with $|I| = i$. Since there are $\binom{N}{i}$ subsets with cardinality i, the formula follows. $\qquad\square$

Corollary 7.
For $m \leq N - 1$, it holds

$$W_j(N) = \sum_{i=1}^{N} (-1)^{i-1} \binom{N}{i} \left(t^{jm^2} |GL_m(\mathbb{F}^j)| \right)^i W_j(N-i).$$

Proof.
If $\det(K_i) = 0$ for $i = 1, \ldots, N$, the column rank of $\mathcal{K}_N$ is at most $Nm - N < Nm - m = (N-1)m$ and therefore, one has no full row rank. Consequently, $\hat{W}_j = 0$ and the statement follows from the preceding theorem. $\qquad\square$

To compute $W_j(N)$ with the help of one of the preceding recursion formulas, one either needs a formula for $\hat{W}_j(N)$ or one has to know $W_j(N)$ for $N \leq m$. In the following, we will take the first option and calculate $\hat{W}_j(N)$. Combining some known results, it is possible to get a conjecture for a formula for $\hat{W}_j(N)$, which we will prove in the following theorem. The stated formula could be motivated in the following way: One could observe that the probability that a vector of length j from $\mathbb{F}$ has no full rank, i.e. is equal to zero, is equal to $1 - t^j$, while for a vector of j polynomials from $\mathbb{F}[z]$, the probability that there exists $z_0 \in \overline{\mathbb{F}}$ such that this vector has no full rank at $z_0 \in \overline{\mathbb{F}}$, i.e. the polynomials are not coprime, is equal to $1 - t^{j-1}$. This additional factor t^{-1} for the subtrahend is somehow due to the possibilities for the choice of z_0. An analogous observation could be made if one compares the probability of pairwise coprimeness for polynomials with the probability for a vector from $\mathbb{F}$ to contain at most one zero entry (see also the matrix $\mathcal{D}_N/(f)$ in the proof of Theorem 17). However, in this case, one only knows that the coefficients of the leading terms in the asymptotic expansion coincide. Therefore, it seems likely that the leading terms of the asymptotic expressions for the probabilities of $\{\det(K_i) = 0 \text{ for } i = 1, \ldots, N\} \setminus \hat{A}$ and $\hat{A}_{N+1}(N)$ defined as in the proof of Theorem 21 differ by the factor t^{-1} as well. Note that in the following formula, one has t^j instead of t because one is dealing with the field $\mathbb{F}^j$.

Lemma 12.
For $j \in \mathbb{N}$ and $N \geq 2$, it holds:

$$\hat{W}_j(N) = \left(1 - t^{jm^2}|GL_m(\mathbb{F}^j)|\right)^N - \sum_{i=N-1}^{\min(m,N-1)} t^{j(m+1)} + O(t^{(m+2)j}).$$

Proof.
Denote by $\tilde{W}$ the probability that $\det(K_i) = 0$ for $i = 1, \ldots, N$ and $\mathcal{K}_N$ is not of full row rank. We will show

$$\tilde{W} = \sum_{i=N-1}^{\min(m,N-1)} t^{(m+1)j} + O(t^{(m+2)j}).$$

The result follows since the sum of $\tilde{W}$ and $\hat{W}_j(N)$ is equal to the probability that $\det(K_i) = 0$ for $i = 1, \ldots, N$.
If $m < N - 1$, the probability that $\det(K_i) = 0$ for $i = 1, \ldots, N$ is equal to $\left(1 - t^{jm^2}|GL_m(\mathbb{F}^j)|\right)^N = (1 - (1 - t^j + O(t^{j+1})))^N = O(t^{jN}) = O(t^{j(m+2)})$, which is conform with $\sum_{i=N-1}^{\min(m,N-1)} t^{j(m+1)} = 0$ in this case.
Next, consider the case $m \geq N - 1$. We have to compute the probability that there exists $\xi \in (\mathbb{F}^j)^{1 \times m(N-1)} \setminus \{0\}$ with $\xi \mathcal{K}_N = 0$, i.e. that there exist $\xi_i \in (\mathbb{F}^j)^{1 \times m}$ for $i = 1, \ldots, N - 1$ which are not all identically zero such that $\xi_1 K_1 = (\xi_1 + \xi_2)K_2 = \cdots = (\xi_{N-2} + \xi_{N-1})K_{N-1} = \xi_{N-1}K_N = 0$.
As in the proof for Remark 11, one could show that either $\xi_i \neq 0$ for $i = 1, \ldots, N - 1$ and $\xi_i + \xi_{i+1} \neq 0$ for $i = 1, \ldots, N - 2$ or there exists $i \in \{1, \ldots, N\}$ such that $\mathcal{K}_{N-1}^{(i)}$ formed by the matrices from the set $\{K_1, \ldots, K_N\} \setminus \{K_i\}$ is not of full row rank. Per induction with respect to N, one knows that the probability for this is $O(t^{j(m+1)})$. Multiplication with the probability that $\det(K_i) = 0$, which is $O(t^j)$, leads to a term for the probability that is $O(t^{j(m+2)})$. Note that one could use induction since for $N = 2$, $\hat{W}$ is just equal to $1 - t^{j \cdot 2m^2} N(2m, m, m) = 1 - \prod_{l=m+1}^{2m}(1 - t^{jl})$ (see Lemma 4) because that $[K_1 \ K_2]$ is not of full row rank already implies $\det(K_1) = \det(K_2) = 0$.

Thus, one could assume $\xi_i \neq 0$ and $\xi_i + \xi_{i+1} \neq 0$.

According to Lemma 4, the probability that $\dim(\ker(K_i)) = r_i$ is equal to

$$t^{jm^2} \cdot N(m, m, m - r_i) = t^{jmr_i} \cdot \prod_{l=r_i+1}^{m} (1 - t^{jl}) \prod_{l=0}^{m-(r_i+1)} \frac{t^{j(l-m)} - 1}{t^{-j(l+1)} - 1} =$$

$$= t^{jmr_i} (1 + O(t^j)) \cdot \frac{\prod_{l=r_i+1}^{m} (t^{-jl} - 1)}{\prod_{l=1}^{m-r_i} (t^{-jl} - 1)} =$$

$$= t^{jmr_i} (1 + O(t^j)) \cdot t^{-\frac{j}{2}(m(m+1) - r_i(r_i+1) - (m-r_i)(m-r_i+1))} =$$

$$= t^{jr_i^2} (1 + O(t^j)).$$

Fix $1 \leq r_i \leq m$ for $i = 1, \ldots, N$. Then, the probability that $\dim(\ker(K_1)) = r_1$ is $t^{jr_1^2} \cdot (1 + O(t^j))$. For each such matrix K_1, there are t^{-jr_1} possibilities for $\xi_1 \in (\mathbb{F}^j)^{1 \times m}$ with $\xi_1 K_1 = 0$. Furthermore, the probability that $\dim(\ker(K_2)) = r_2$ is $t^{jr_2^2} \cdot (1 + O(t^j))$ and for fixed ξ_1 and K_2, there are t^{-jr_2} possibilities for $\xi_2 \in (\mathbb{F}^j)^{1 \times m}$ such that $(\xi_1 + \xi_2)K_2 = 0$. This procedure is continued until K_i and ξ_i are fixed for $i = 1, \ldots, N - 1$. As we assumed $\xi_{N-1} \neq 0$, the probability for K_N to fulfil $\xi_{N-1} K_N = 0$ is equal to t^{jm}.

Finally, one has to consider, which values for $\xi_1, \ldots, \xi_{N-1}$ lead to the same solutions for $K_1, \ldots, K_N$. One clearly gets the same solutions if one multiplies ξ_i for $i = 1, \ldots, N - 1$ by the same scalar value, which effects a factor that is $O(t^j)$ for the probability. In summary, the overall probability is $O\left(t^{j(\sum_{i=1}^{N-1}(r_i^2 - r_i) + m + 1)}\right)(1 + O(t^j))$. Hence, all cases in which $r_i \geq 2$ for some $i \in \{1, \ldots, N - 1\}$ could be neglected.

It remains to show that for $r_1 = \cdots = r_N = 1$, only $\xi_1, \ldots, \xi_{N-1}$ which differ all by the same scalar factor lead to the same solutions for $K_1, \ldots, K_N$. Then, one knows that the factor for the probability caused by this effect is exactly t^j and one gets a overall probability of $t^{(m+1)j} + O(t^{(m+2)j})$, which is conform with $\sum_{i=N-1}^{\min(m,N-1)} t^{j(m+1)} = t^{j(m+1)}$ in the considered case $m \geq N - 1$.

To do this, we firstly show that the case that $\xi_1, \ldots, \xi_{N-1}$ are linearly dependent could be neglected. For the choice of such vectors ξ_i with the property that $\text{rk}[\xi_1^\top \cdots \xi_{N-1}^\top] < N - 1$, one has

$$O\left(\sum_{r=1}^{N-2} N(m, N - 1, r)\right) = O\left(\sum_{r=1}^{N-2} t^{-jr(m+N-1-r)}\right) = O(t^{-j(N-2)(m+1)})$$

possibilities and for each of these possibilities the probability that $\xi_1 K_1 = (\xi_1 + \xi_2)K_2 = \cdots = (\xi_{N-2} + \xi_{N-1})K_{N-1} = \xi_{N-1} K_N = 0$ is equal to t^{jNm} as $\xi_i \neq 0$ and $\xi_i + \xi_{i+1} \neq 0$. Additionally, one has again a factor of $O(t^j)$ because of the values for the vectors ξ_i that lead to the same solutions for $K_1, \ldots, K_N$. In summary, one gets a probability that is $O(t^{j(Nm+1-(N-2)(m+1))}) = O(t^{j(m+2)})$ since $-N \geq -m - 1$.

Hence, in the following, one could assume that $\xi_1, \ldots, \xi_{N-1}$ are linearly independent. If $\xi_1 K_1 = \tilde{\xi}_1 K_1 = 0, (\xi_1 + \xi_2)K_2 = (\tilde{\xi}_1 + \tilde{\xi}_2)K_2 = 0, \ldots, \xi_{N-1} K_N = \tilde{\xi}_{N-1} K_N = 0$, it results from $r_1 = \cdots = r_N = 1$ that there exist $\lambda_i \in \mathbb{F}^j$ with $\tilde{\xi}_1 = \lambda_1 \xi_1, \tilde{\xi}_i + \tilde{\xi}_{i+1} = \lambda_{i+1}(\xi_i + \xi_{i+1})$ for $i = 1, \ldots, N - 2$ and $\tilde{\xi}_{N-1} = \lambda_N \xi_{N-1}$. Since $\tilde{\xi}_1 - (\tilde{\xi}_1 + \tilde{\xi}_2) + \cdots \pm (\tilde{\xi}_{N-2} + \tilde{\xi}_{N-1}) \mp \tilde{\xi}_{N-1} = 0$, it follows

$(\lambda_1 - \lambda_2)\xi_1 + (\lambda_3 - \lambda_2)\xi_2 + \cdots \pm (\lambda_{N-1} - \lambda_N)\xi_{N-1} = 0$. As $\xi_1, \ldots, \xi_{N-1}$ are linearly independent, this implies $\lambda_1 = \cdots = \lambda_N$, which completes the proof of the whole theorem. □

With this result, we are able to solve the recursion formula from Lemma 11 and obtain an explicit expression for $W_j(N)$.

Theorem 24.
For $j \in \mathbb{N}$ and $N \geq 2$, the probability that N matrices from $(\mathbb{F}^j)^{m \times m}$ are mutually left coprime is equal to

$$W_j(N) = 1 - \sum_{y=2}^{m+1} \binom{N}{y} t^{j(m+1)} + O(t^{j(m+2)}).$$

Proof.
This is shown per induction with respect to N.
For $N = 2$, one just has to compute the probability that a rectangular matrix is of full rank. According to Lemma 4 with $n = 2m$ and $k = r = m$, this probability is equal to $\prod_{i=m+1}^{2m}(1 - (t^j)^i) = 1 - t^{j(m+1)} + O(t^{j(m+2)})$.
Inserting the assumption of the induction into the first part of the recursion formula from Lemma 11, leads to

$$\sum_{i=1}^{N}(-1)^{i-1}\binom{N}{i}\left(t^{jm^2}|GL_m(\mathbb{F}^j)|\right)^i W_j(N-i) =$$

$$= \sum_{i=1}^{N}(-1)^{i-1}\binom{N}{i}\left(t^{jm^2}|GL_m(\mathbb{F}^j)|\right)^i \left(1 - \sum_{y=2}^{m+1}\binom{N-i}{y}t^{j(m+1)} + O(t^{j(m+2)})\right)$$

$$= \sum_{i=1}^{N}(-1)\binom{N}{i}\left((-1)t^{jm^2}|GL_m(\mathbb{F}^j)|\right)^i +$$

$$+ \sum_{i=1}^{N}(-1)^i\binom{N}{i}\sum_{y=2}^{m+1}\binom{N-i}{y}t^{j(m+1)} + O(t^{j(m+2)}) =$$

$$= -\left(1 - t^{jm^2}|GL_m(\mathbb{F}^j)|\right)^N + 1 +$$

$$+ \sum_{i=1}^{N-2}(-1)^i\binom{N}{i}\sum_{y=2}^{m+1}\binom{N-i}{y}t^{j(m+1)} + O(t^{j(m+2)}) =$$

$$= -\left(1 - t^{jm^2}|GL_m(\mathbb{F}^j)|\right)^N + 1 - \sum_{y=2}^{\min(m+1,N-1)}\binom{N}{y}t^{j(m+1)} + O(t^{j(m+2)}).$$

The final step of the preceding computation was done analogous to the proof of Remark 13. Using the formula for $\hat{W}_j(N)$ from the preceding theorem, one obtains

$$W_j(N) = -\left(1 - t^{jm^2}|GL_m(\mathbb{F}^j)|\right)^N + 1 - \sum_{y=2}^{\min(m+1,N-1)} \binom{N}{y} t^{j(m+1)} + \hat{W}_j(N)$$

$$= 1 - \left(\sum_{i=N-1}^{\min(m,N-1)} 1 + \sum_{y=2}^{\min(m+1,N-1)} \binom{N}{y}\right) t^{j(m+1)} + O(t^{j(m+2)}) =$$

$$= 1 - \sum_{y=2}^{m+1} \binom{N}{y} t^{j(m+1)} + O(t^{j(m+2)}).$$

For the last equality, see end of the proof for Remark 13. □

Finally, connecting the preceding results, one gets the following formula for the natural density of mutual left coprimeness.

Theorem 25.
The natural density of N matrices $D_i \in \mathbb{F}[z]^{m \times m}$ for $i = 1, \ldots, N$ to be mutually left coprime is equal to

$$\prod_{j=1}^{\infty} \left(1 - \sum_{y=2}^{m+1} \binom{N}{y} t^{j(m+1)} + O(t^{j(m+2)})\right)^{\varphi_j} = 1 - \sum_{y=2}^{m+1} \binom{N}{y} t^m + O(t^{m+1}).$$

2.5 Conclusion

In the preceding chapter, we firstly computed the probabilities for several coprimeness properties using the uniform probability distribution, i.e. bounding the degrees of the polynomial entries of the involved matrices. Afterwards, we compared these results with the formula one gets using the natural density, i.e. allowing the degrees of the corresponding polynomials to grow arbitrarily large. It is remarkable that uniform probability and natural density have asymptotically the same values in all computed cases, while one could observe differences in some of the calculated exact formulas. In the following section, we will apply the achieved results to the calculation of the probabilities that networks of linear systems are reachable and/or observable, respectively. Since the degrees of the entries of the corresponding polynomial matrix fraction descriptions are bounded by the parameters of the linear systems, this could only be done using the uniform probability distribution. Consequently, we will not employ the natural density for our further considerations.

Chapter 3

Probabilities of Reachability and Observability for Interconnected Linear Systems

In this chapter, we will use the results of the preceding chapters to compute or estimate, the probabilities that certain networks of linear systems are reachable, observable or minimal.

We start in Section 3.1 with the simplest "network", namely just one system. For this, the probability of reachability and therefore, also the probability of observability are already known. We recall these results and moreover, compute the probability that a linear system is minimal.

Section 3.2 deals with parallel connection of linear systems. To obtain a formula for the probability of reachability and observability, one mainly needs the probability of mutual left coprimeness from the preceding chapter.

In Section 3.3, we consider series connections of linear systems, firstly for single-input single-output node systems and secondly, for two systems with arbitrary parameters. Finally, we investigate how the corresponding probabilities of reachability are changed if one requires the transfer functions of the node systems to be strictly proper.

Section 3.4 is similarly structured as Section 3.3 but deals with circular interconnection. Since series and circular interconnection are quite similar, several computations carry over. However, some adjustments are necessary.

In Section 3.5, we consider again series connections and investigate their so-called $(1,3)$-reachability, i.e. for some special values for the parameters of the involved systems, we examine criteria for the possibility to control only the first and the third system of the series.

In contrast to the preceding sections, Section 3.6 considers no special typ of interconnection but is devoted to the question what general statements about the corresponding probability are possible without knowing the exact interconnection structure of a network.

Finally, in Section 3.7, we consider so-called homogeneous networks and calculate the probability that these are reachable or/and observable if the matrices defining the interconnection structure are chosen randomly.

3.1 Probabilities for a Single System

3.1.1 Reachability and Observability

The probability of reachability for a single system has already been computed in [21]. We will need it for computing the probability of minimality for a single system as well as for the calculation of probabilities for interconnections of $N \geq 2$ systems since reachability and observability of the node systems are part of the criteria for reachability and observability of the whole network; see Corollary 1.

Theorem 26. *[21, Theorem 1]*
Let $\Sigma_{n,m}^{cr}(\mathbb{F})$ denote the set of all reachable pairs. The probability that a pair $(A, B) \in \mathbb{F}^{n \times n} \times \mathbb{F}^{n \times m}$ is reachable is equal to

$$P_{n,m}(t) := \frac{|\Sigma_{n,m}^{cr}(\mathbb{F})|}{|\mathbb{F}^{n \times n} \times \mathbb{F}^{n \times m}|} = \prod_{j=m}^{n+m-1} (1 - t^j) = 1 - t^m + O(t^{m+1}). \tag{3.1}$$

In particular, one obtains for $n \geq 2$:

$$(1 - t^m)(1 - (n-1)t^{m+1}) \leq P_{n,m}(t) \leq (1 - t^m)(1 - t^{m+1}). \tag{3.2}$$

Using Remark 1, one could easily deduce the probability of observability:

Corollary 8.
The probability that a pair $(A, C) \in \mathbb{F}^{n \times n} \times \mathbb{F}^{p \times n}$ is observable is equal to

$$\prod_{j=p}^{n+p-1} (1 - t^j) = 1 - t^p + O(t^{p+1}).$$

Since the number of reachable pairs is strongly connected with the number of matrices in Hermite form, Theorem 26 leads to the following corollary as well:

Corollary 9.
The number of nonsingular polynomial matrices $Q \in \mathbb{F}[z]^{m \times m}$ in Hermite form whose determinant is a (monic) polynomial of degree n is equal to

$$H_{n,m}(\mathbb{F}) = \sum_{\kappa_1 + \cdots + \kappa_m = n} t^{-\sum_{i=1}^{m}(m-i+1)\cdot\kappa_i} = \frac{|\Sigma_{n,m}^{cr}(\mathbb{F})|}{|GL_n(\mathbb{F})|} = t^{-mn} \prod_{j=1}^{n} \frac{1 - t^{(m+j-1)}}{1 - t^j}. \tag{3.3}$$

Proof.
The statement of the first equation is already contained in Lemma 2.1 and the second equation of (3.3) follows from the proof of Theorem 1 of [21]. Finally, the third equation is a consequence of this theorem itself, i.e. of Theorem 26 of this paper, and of Theorem 4. □

Remark 14.
Since both Hermite form as well as Kronecker-Hermite form are unique, the number of Kronecker-Hermite forms is equal to $H_{n,m}(\mathbb{F})$, as well.

3.1.2 Minimality

The aim of this section is to compute the probability that a linear system is both reachable as well as observable. Since these two properties are not independent of each other (see Remark 17), it is not possible to simply multiply the probability of reachability with the probability of observability. However, the following lemma states that the probability of minimality could be obtained by multiplying the probability of reachability with the probability of right primeness.

Lemma 13.
The probability that a linear discrete-time system described by $(A, B, C, D) \in \mathbb{F}^{n \times n} \times \mathbb{F}^{n \times m} \times \mathbb{F}^{p \times n} \times \mathbb{F}^{p \times m}$ is minimal is equal to $P_{p,n,m}^{rc}(t) \cdot P_{n,m}(t)$.

Proof.
For each $(A, B, C, D) \in \mathbb{F}^{n \times n} \times \mathbb{F}^{n \times m} \times \mathbb{F}^{p \times n} \times \mathbb{F}^{p \times m}$ describing a minimal system, there exists exactly one right coprime pair $(P, Q) \in \mathbb{F}[z]^{p \times m} \times \mathbb{F}[z]^{m \times m}$ where Q is in Kronecker-Hermite form and $C(zI - A)^{-1}B + D = P(z)Q(z)^{-1}$. According to Theorem 9 (a), it holds $\deg(\det(Q)) = n$ and according to Lemma 1, $\deg_j P(z) \leq \deg_j Q(z)$ for $j = 1, \ldots, m$. On the other hand, for every such pair (P, Q), there exist exactly $|GL_n(\mathbb{F})|$ minimal realizations (A, B, C, D); see Theorem 4. Consequently, the number of minimal systems is equal to the number of pairs (P, Q) times $|GL_n(\mathbb{F})|$. According to Remark 14, the number of Kronecker-Hermite forms is equal to $\frac{|\Sigma_{n,m}^{cr}(\mathbb{F})|}{|GL_n(\mathbb{F})|}$. Moreover, for each of them, there are $\prod_{i=1}^{m} t^{-p(\kappa_i+1)} = t^{-p\sum_{i=1}^{m}(\kappa_i+1)} = t^{-p(n+m)}$ polynomial matrices $P \in \mathbb{F}[z]^{p \times m}$ which fulfill $\deg_j P(z) \leq \deg_j Q(z)$ for $j = 1, \ldots, m$. Consequently, the corresponding probability is equal to

$$\frac{P_{p,n,m}^{rc}(t) \cdot t^{-(np+mp)} \cdot \frac{|\Sigma_{n,m}^{cr}(\mathbb{F})|}{|GL_n(\mathbb{F})|} \cdot |GL_n(\mathbb{F})|}{|\mathbb{F}^{n \times n} \times \mathbb{F}^{n \times m} \times \mathbb{F}^{p \times n} \times \mathbb{F}^{p \times m}|} = \frac{P_{p,n,m}^{rc}(t) \cdot |\Sigma_{n,m}^{cr}(\mathbb{F})|}{|\mathbb{F}^{n \times n} \times \mathbb{F}^{n \times m}|} =$$

$$= P_{p,n,m}^{rc}(t) \cdot P_{n,m}(t).$$

$\square$

Remark 15.
This formula is also valid if one considers strictly proper transfer functions, i. e. if one sets $D = 0$. In this case, there are t^{-pm} possibilities less for P, which equals the possibilities for D. Since minimality of a system does not depend on D, one could conclude that the probability of right primeness is independent of D as well (see also Remark 8).

Remark 16.
Since a linear system is minimal if and only if it is reachable and observable, it follows from the preceding lemma that the probability that a system is observable under the condition that it is reachable is equal to $P_{p,n,m}^{rc}(t) = 1 - t^p + O(t^{p+1})$. This means that the ratio of observable systems among all systems is approximately the same as the ratio of observable systems among reachable systems. However, the following remark will show that this is, in general, only asymptotically true but not for the exact probability values.

With the help of Theorem 13 it is easy to compute the exact probability of minimality for the case of single-input single-output systems:

Corollary 10. *[21]*
The probability that a single-input single-output system $(A, B, C, D) \in \mathbb{F}^{n \times n} \times \mathbb{F}^{n \times 1} \times \mathbb{F}^{1 \times n} \times \mathbb{F}$ is minimal is equal to

$$\prod_{j=1}^{n}(1 - t^j)(1 - t).$$

Proof.
Since $p = m = 1$, one has the exact probability $1 - t$ for the scalar polynomials P and Q to be coprime. □

Remark 17.
The case $p = m = 1$ shows that, in general, reachability and observability are not independent properties of a linear system. If this would be true, the probability of minimality had to be equal to the product of the probabilities for reachability and observability, which is $\prod_{j=1}^{n}(1 - t^j) \cdot \prod_{j=1}^{n}(1 - t^j)$ for the case of single input. But this is only true for $n = 1$. However, in this case, it is clear that the two properties are independent because then, reachability is equivalent to $B \neq 0$ and observability is equivalent to $C \neq 0$; note that in this case B and C are scalar.

To achieve an asymptotic result for the probability of minimality of a linear system with arbitrary parameters, one has to use the asymptotic formula for right primeness from Theorem 12 and insert it into the expression from Theorem 13.

Theorem 27.
The probability that a linear discrete-time system described by $(A, B, C, D) \in \mathbb{F}^{n \times n} \times \mathbb{F}^{n \times m} \times \mathbb{F}^{p \times n} \times \mathbb{F}^{p \times m}$ is minimal is equal to

$$1 - t^m - t^p + O(t^{\min(p,m)+1}).$$

3.2 Parallel Connection

The aim of this section is to compute the probability that the parallel connected system

$$x_1(\tau + 1) = A_1 x_1(\tau) + B_1 u(\tau)$$

$$\vdots \tag{3.4}$$

$$x_N(\tau + 1) = A_N x_N(\tau) + B_N u(\tau)$$

with state vectors $x_i \in \mathbb{F}^{n_i}$ for $i = 1, \dots, N$ and input $u \in \mathbb{F}^m$ is reachable.

System (3.4) contains no outputs and therefore, could be interpreted as a special case of parallel connection (as defined in Example 1 (a)) with $C = I$ and $D = 0$, which implies the observability of the node systems. Thus, reachability of the node systems, which is necessary for the reachability of the interconnected system according to Remark 4, is equivalent to the minimality of the node systems, which is the condition for applying Theorem 10. Consequently, reachability of the interconnection is equivalent to the reachability of the node systems together with the corresponding coprimeness condition of Example 1 (a). Our aim is to count the number of reachable interconnections by counting possible coprime factorizations of the transfer functions of the node systems. Therefore, we need the following statement.

Lemma 14.
Let $Q \in \mathbb{F}[z]^{m \times m}$ nonsingular be in Hermite form with $\deg(\det(Q(z))) = n$. Then, there are exactly $|GL_n(\mathbb{F})|$ reachable pairs $(A, B) \in \mathbb{F}^{n \times n} \times \mathbb{F}^{n \times m}$ with $(zI - A)^{-1}B = P(z)Q(z)^{-1}$ for some $P \in \mathbb{F}[z]^{n \times m}$ such that P and Q are right coprime. In other words, there are exactly $|GL_n(\mathbb{F})|$ polynomial matrices $P \in \mathbb{F}[z]^{n \times m}$ such that P and Q are right coprime and PQ^{-1} could be written in the form $P(z)Q(z)^{-1} = (zI - A)^{-1}B$, where $(A, B) \in \mathbb{F}^{n \times n} \times \mathbb{F}^{n \times m}$ is reachable.

Proof.
According to Proposition 2.3 of [43], there exist a reachable pair (A, B) and a polynomial matrix P that is right coprime to Q, such that $(zI - A)^{-1}B = P(z)Q(z)^{-1}$. Now, one considers the orbit of this pair (A, B) under the similarity action on the state space, i.e. the set $\{(TAT^{-1}, TB) \mid T \in GL_n(\mathbb{F})\}$, which clearly consists only of reachable pairs. If $(zI - TAT^{-1})^{-1}TB = \tilde{P}(z)\tilde{Q}(z)^{-1}$ is a right coprime factorization of the transfer function with $\tilde{Q}$ in Hermite form, it follows from Theorem 2.4 a, of [43] that $Q = \tilde{Q}U$ with a unimodular matrix $U \in GL_n(\mathbb{F}[z])$. But since the Hermite form of a matrix is unique and $\tilde{Q}$ and Q are both in Hermite form, one knows $\tilde{Q} = Q$. Thus, Q leads to at least $|GL_n(\mathbb{F})|$ reachable realizations (A, B). On the other hand, the reverse direction of the statement of Theorem 2.4 a, of [43] shows that the right coprime factorizations $(zI - A_1)^{-1}B_1 = P_1(z)Q(z)^{-1}$ and $(zI - A_2)^{-1}B_2 = P_2(z)Q(z)^{-1}$ together with the reachability of (A_1, B_1) and (A_2, B_2) imply $(A_2, B_2) = (TA_1T^{-1}, TB_1)$ for some $T \in GL_n(\mathbb{F})$. Therefore, Q leads to at most $|GL_n(\mathbb{F})|$ reachable realizations (A, B). $\square$

The following lemma shows that the two conditions, which have to be considered for parallel connection, i.e. the reachability of the node systems and the mutual left coprimeness of the matrices $Q_1, \ldots, Q_N$, are in fact independent.

Lemma 15.
Let $(A, B) \in \mathbb{F}^{n \times n} \times \mathbb{F}^{n \times m}$ and $G(z) = (zI - A)^{-1}B = P(z)Q(z)^{-1}$ be the corresponding transfer function with $P \in \mathbb{F}[z]^{n \times m}, Q \in \mathbb{F}[z]^{m \times m}$, where $\det(Q) \neq 0$. Then, the reachability of (A, B) only depends on P.

Proof.
By the Kalman test (see Theorem 1), system (A, B) is reachable if and only if $c = 0$ is the only solution of $cA^iB = 0$ for $0 \leq i \leq n - 1$ with $c^\top \in \mathbb{F}^n$. Note that $cA^iB = 0$ for $0 \leq i \leq n - 1$ implies $cA^iB = 0$ for $i \geq 0$ by the theorem of Cayley-Hamilton. Since $(zI - A)^{-1} = \sum_{i=0}^{\infty} \frac{A^i}{z^{i+1}}$, reachability is equivalent to the fact that $c = 0$ is the

only solution of $c(zI - A)^{-1}B \equiv 0$ with $c^\top \in \mathbb{F}^n$. This means that $cP \equiv 0$ for $c^T \in \mathbb{F}^n$ implies $c = 0$, which is a criterion that only depends on P. □

As a consequence of this result, one could just multiply the probabilities of the two conditions mentioned before and thus, we are ready to prove the main theorem of this section, providing a formula for the probability of reachability for a parallel connected system.

Theorem 28.
The probability that the parallel connected system given by (3.4) is reachable is

$$\prod_{i=1}^{N} \prod_{j=m}^{n_i+m-1} (1 - t^j) \cdot P_m(N),$$

where $P_m(N)$ is the probability that N polynomial matrices from $\mathbb{F}[z]^{m \times m}$ in Hermite form are mutually left coprime.

Proof.
Consider the right coprime factorizations $(zI - A_i)^{-1}B_i = P_i(z)Q_i(z)^{-1}$ for $i = 1, \ldots, N$. From Theorem 9 (a) one knows that $\deg(\det(Q_i)) = n_i$ for $i = 1, \ldots, N$ and from Theorem 9 (b) that one could assume that the polynomial matrices $Q_1, \ldots, Q_N$ are in Hermite form. According to Lemma 14, for each such Q_i, there exists the same number of reachable pairs (A_i, B_i), namely exactly $|GL_{n_i}(\mathbb{F})|$. Therefore, the probability that the coprimeness condition of Example 1 (a) for reachability is fulfilled is equal to the probability that arbitrary polynomial matrices Q_i (in Hermite form) with $\deg(\det(Q_i)) = n_i$ for $i = 1, \ldots, N$ are mutually left coprime. Since this condition only depends on the matrices Q_i and according to Lemma 15, the reachability of the node systems only depends on P_i, one could just multiply the probability of mutual left coprimeness with the probabilities that the node systems are reachable (see Theorem 26 for the corresponding formula). □

Remark 18.
Since for parallel connection it holds $K = 0$ and hence, A and B are independent of (C_i, D_i) for $i = 1, \ldots, N$, the formula of the preceding theorem is also valid if (C_i, D_i) are chosen randomly for $i = 1, \ldots, N$ (and are not fixed to $(I, 0)$ as done there).

For $m = 1$, i.e. in the case of single-input, one could combine Theorem 28 with Theorem 15 to get a formula for the probability of reachability for a parallel connection.

Theorem 29.
The probability that a parallel connection of N single-input systems is reachable is equal to

$$\prod_{l=1}^{N} \left(\prod_{j=1}^{n_l} (1 - t^j) \right) \sum_{k \in M(n)} \prod_{ij \in \mathcal{E}} (-1)^{\omega(k_{ij})} \prod_{l=1}^{N} t^{\deg(K_l)},$$

where $\mathcal{E} = \{ij \mid i, j \in \{1, \ldots, N\}, i < j\}$.

To get an estimation for this quite complicated formula, we will we calculate its asymptotic behaviour by using Theorem 16.

Theorem 30.
The probability that a parallel connection of N single input systems is reachable is equal to

$$1 - \frac{N(N+1)}{2} \cdot t +$$
$$+ \left(\frac{1}{24}(N-1)(N-2)(3N^2 + 11N - 12N_1) + \frac{N^3 - 3N}{2} \right) \cdot t^2 + O(t^3),$$

where N_1 is the number of systems whose state vectors are scalar.

Proof.
Since the probability that the single systems are reachable is equal to
$\prod_{j=1}^{N} \prod_{i=1}^{n_j}(1 - t^i) = 1 - N \cdot t + \left(\frac{N(N-1)}{2} - N \right) \cdot t^2 + O(t^3)$, one obtains

$$\left(1 - \frac{N(N-1)}{2} \cdot t + \frac{1}{24}(N-1)(N-2)(3N^2 + 11N - 12N_1) \cdot t^2 + O(t^3) \right) \cdot$$
$$\cdot \left(1 - N \cdot t + \left(\frac{N(N-1)}{2} - N \right) \cdot t^2 + O(t^3) \right)$$

for the probability that the parallel connection is reachable. Computing the coefficient of t^2, results in

$$\frac{1}{24}(N-1)(N-2)(3N^2 + 11N - 12N_1) + \frac{N(N-1)}{2} \cdot N + \frac{N(N-1)}{2} - N =$$
$$= \frac{1}{24}(N-1)(N-2)(3N^2 + 11N - 12N_1) + \frac{N^3 - 3N}{2}.$$

$\square$

Furthermore, it is possible to obtain an asymptotic formula for general $m \in \mathbb{N}$, including the coefficients of $1, \ldots, t^m$ by using Theorem 22.

Theorem 31.
The probability that the parallel connection of N linear systems with m inputs is reachable is equal to

$$1 - \sum_{y=1}^{m+1} \binom{N}{y} t^m + O(t^{m+1}).$$

Proof.
Combining the already achieved results, leads to

$$\prod_{i=1}^{N} \prod_{j=m}^{n_i+m-1} (1-t^j) \cdot P_m(N) =$$

$$= (1 - N \cdot t^m + O(t^{m+1})) \cdot \left(1 - \sum_{y=2}^{m+1} \binom{N}{y} t^m + O(t^{m+1})\right) =$$

$$= 1 - \sum_{y=1}^{m+1} \binom{N}{y} t^m + O(t^{m+1}).$$

$\square$

If one considers the formula of the preceding theorem, one could observe that the modulus of the coefficient of t^m is increasing with N as well as with m. Thus, the probability decreases with N and increases with t^{-1}. No general statement is possible how increasing m influences the term $\sum_{y=1}^{m+1} \binom{N}{y} t^m$. However, one could see that it decreases with m (and hence, the probability increases) if $m \geq N$ or if t tends to zero, which could be assumed since the above formula is only valid for $t^{-1} \to \infty$. Using the duality between reachability and observability, one gets the folllowing corollary:

Corollary 11.
For $B = I$ and $D = 0$ (with the notation of Theorem 10), the probability that the parallel connection of N systems with p outputs is observable is equal to $1 - \sum_{y=1}^{p+1} \binom{N}{y} t^p + O(t^{p+1})$.

Proof.
The proof is analogue to that for the reachability of a parallel connection but the roles of m and p are interchanged. One applies Theorem 10 (2.) and uses the criterion on $\hat{Q}$ and $\hat{P}$ from Example 1 (a), i.e. one has to calculate the probability

that $\begin{bmatrix} \hat{Q}_1 & & & 0 \\ \hat{Q}_2 & \hat{Q}_2 & & \\ & \ddots & \ddots & \\ 0 & & \hat{Q}_N & \hat{Q}_N \end{bmatrix}$ is right prime, which is equivalent to the fact that

$$\begin{bmatrix} \hat{Q}_1 & & & 0 \\ \hat{Q}_2 & \hat{Q}_2 & & \\ & \ddots & \ddots & \\ 0 & & \hat{Q}_N & \hat{Q}_N \end{bmatrix}^{\mathsf{T}} = \begin{bmatrix} \hat{Q}_1^{\mathsf{T}} & \hat{Q}_2^{\mathsf{T}} & & 0 \\ & \ddots & \ddots & \\ 0 & & \hat{Q}_{N-1}^{\mathsf{T}} & \hat{Q}_N^{\mathsf{T}} \end{bmatrix}$$

is left prime. This in turn is equivalent to the mutual left coprimeness of $\hat{Q}_1^{\mathsf{T}}, \ldots, \hat{Q}_N^{\mathsf{T}} \in \mathbb{F}[z]^{p \times p}$. Note that $\hat{Q}_i^{-1} \hat{P}_i$ is a left coprime factorization of the transfer function G_i

if and only if $\hat{P}_i^\top (\hat{Q}_i^\top)^{-1}$ is a right coprime factorization of G_i^T. Moreover, since $B = I$, the node systems are reachable and similiar to the considerations concerning reachability, one could show that the observability of the node systems only depends on $\hat{P}$. Consequently, one has to multiply the probabilities of observability for the single systems with the probability of mutual left coprimeness, which gives the stated result. $\qquad\qquad\square$

The preceding corollary could also be proven by using directly the duality between observability and reachability. According to Remark 1, $(\mathcal{A}, \mathcal{B}, \mathcal{C})$ is observable if and only if $(\mathcal{A}^\top, \mathcal{C}^\top, \mathcal{B}^\top)$ is reachable. Since for parallel connection, it holds $\mathcal{A} = A$, $\mathcal{B} = (B_1^\top, \ldots, B_N^\top)^\top$ and $\mathcal{C} = (C_1, \ldots, C_N)$, the observability of the parallel connection from the preceding corollary is equivalent to the reachability of the parallel connection from (3.4) with $(A_i^\top, C_i^\top)$ for $i = 1, \ldots, N$ as node systems. Therefore, the formula is also valid if B and D are chosen randomly and one gets the following corollary.

Corollary 12.
The probability that a parallel connection of N systems with p outputs is observable is equal to $1 - \sum_{y=1}^{p+1} \binom{N}{y} t^p + O(t^{p+1})$.

3.3 Series Connection

In this section, we consider series connections as in Example 1 (b). According to this example and to Theorem 10, one has to compute the probability that

$$
\mathcal{R}_N := \begin{bmatrix} P_1 & Q_2 & 0 & \cdots & 0 \\ 0 & P_2 & Q_3 & \ddots & \vdots \\ \vdots & \ddots & \ddots & \ddots & 0 \\ 0 & \cdots & 0 & P_{N-1} & Q_N \end{bmatrix}
$$

is left prime under the condition that the node systems are minimal to get the probability of reachability for the whole interconnection. In contrast to the preceding section, where we studied parallel connections, the above coprimeness condition also depends on the matrices P_i and not only on the matrices Q_i. This means that one has to deal with conditional probability. To this end, define $A_i := \{(P_i, Q_i) \text{ not right coprime}\}$ for $i = 1, \ldots, N$. Moreover, set $A_{N+1} := \{\mathcal{R}_N \text{ not left prime}\}$ and $A_{ij} := A_i \cap A_j$. Then, according to the proof of Lemma 13, the probability that the series connection of N minimal systems is reachable is equal to

$$
\frac{1 - \Pr(\cup_{i=1}^{N+1} A_i)}{1 - \Pr(\cup_{i=1}^{N} A_i)} = \frac{1 - \sum_{\emptyset \neq I \subset \{1,\ldots,N+1\}} (-1)^{|I|-1} \Pr(A_I)}{1 - \sum_{\emptyset \neq I \subset \{1,\ldots,N\}} (-1)^{|I|-1} \Pr(A_I)} =
$$

$$
= 1 - \frac{\Pr(A_{N+1}) + O(\sum_{i=1}^{N} \Pr(A_{i,N+1}))}{1 - \sum_{\emptyset \neq I \subset \{1,\ldots,N\}} (-1)^{|I|-1} \Pr(A_I)}. \tag{3.5}
$$

For the first equality, the inclusion-exclusion principle was used. Moreover, note that for $1 \leq i < j \leq N$, $\Pr(A_{ij}) = \Pr(A_i) \cdot \Pr(A_j)$ because A_i and A_j are independent sets for $1 \leq i < j \leq N$.

3.3.1 SISO-Systems

We start considering a series connection of single-input single-output (SISO) systems, that means $m_i = p_i = 1$ and the coprime factors P_i and Q_i of the transfer functions are scalar for $i = 1, \ldots, N$. Therefore, one has to compute the probability that there exists no $z_0 \in \mathbb{F}$ such that

$$
\mathrm{rk} \begin{bmatrix} P_1 & Q_2 & 0 & \cdots & 0 \\ 0 & P_2 & Q_3 & \ddots & \vdots \\ \vdots & \ddots & \ddots & \ddots & 0 \\ 0 & \cdots & 0 & P_{N-1} & Q_N \end{bmatrix} (z_0) < N - 1.
$$

Doing this, one achieves the following theorem:

Theorem 32.
The probability that the series connection of N minimal single-input single-output systems is reachable is equal to

$$
1 - \frac{N(N-1)}{2} t + O(t^2).
$$

Proof.
One starts computing A_{N+1} and shows per induction with respect to N that $\mathcal{R}_N(z_0)$ is singular if and only if there exist $i \in \{2, \ldots, N\}$ and $j \in \{1, \ldots, i-1\}$ such that $Q_i(z_0) = P_j(z_0) = 0$. The base clause $N = 2$ is trivial.
Now, assume that $\mathcal{R}_{N+1}(z_0)$ is singular and distinguish the cases $Q_{N+1}(z_0) = 0$ and $Q_{N+1}(z_0) \neq 0$. If $Q_{N+1}(z_0) = 0$, $\mathcal{R}_{N+1}(z_0)$ is singular if and only if at least one of the polynomials P_i for $i = 1, \ldots, N$ has a zero at z_0 and we are done. If $Q_{N+1}(z_0) \neq 0$,
$\mathcal{R}_{N+1}(z_0)$ is singular if and only if the matrix $\begin{bmatrix} P_1 & Q_2 & & 0 \\ & \ddots & \ddots & \\ 0 & & P_{N-1} & Q_N \end{bmatrix} (z_0)$ is singular and the statement follows from the induction hypothesis.
As done at the beginning of Section 2.3, regard the polynomials $Q_1, \ldots, Q_N, P_1, \ldots, P_N$ as vertices of a graph Γ. P_i and Q_j are connected by an edge if and only if $j > i$ (there are no edges between two polynomials Q_k and Q_l or P_k and P_l for $k, l \in \{1, \ldots, N\}$, respectively). Thus, Γ has $\sum_{i=1}^{N-1}(N-i) = \sum_{i=1}^{N-1} i = \frac{N(N-1)}{2}$ edges. Using Corollary 3 instead of Lemma 7 (a), one could argue as in the proof of Theorem 14 and gets $\Pr(A_{N+1}) = \frac{N(N-1)}{2} t + O(t^2)$.
We already know $\Pr(A_i) = t + O(t^2)$ and therefore, $\Pr(A_{ij}) = \Pr(A_i) \cdot \Pr(A_j) = O(t^2)$ for $i, j \leq N$. It remains to show $\Pr(A_{i,N+1}) = O(t^2)$ for $1 \leq i \leq N$. But this is equal to the probability that P_i and Q_i are not coprime and that there exist $j < k$ such that

Q_k and P_j are not coprime. For $i \notin \{k, j\}$, it is clear that this probability is $O(t^2)$. If $i = k$, one additionally needs Lemma 7 (a), and if $i = j$, Corollary 3 to get that this probability is $O(t^2)$. Inserting all achieved results into (3.5), leads to a probability for reachability of $1 - \frac{\frac{N(N-1)}{2}t + O(t^2)}{1 + O(t)} = 1 - \frac{N(N-1)}{2}t + O(t^2)$. $\qquad\square$

Remark 19.
If $P_i = Q_i$ for $i = 1, \dots, N$, the preceding primality criterion would correspond to pairwise coprimeness of $Q_1, \dots, Q_N$.

Using again the duality between reachability and observability, one could deduce the probability of observability.

Corollary 13.
The probability that the series connection of N minimal single-input single-output systems is observable is equal to

$$1 - \frac{N(N-1)}{2}t + O(t^2).$$

In contrast to parallel connection, for series connection, the probability of reachability changes if one requires the transfer functions of the node systems to be strictly proper.

Theorem 33.
The probability that the series connection of N minimal single-input single-output systems with strictly proper transfer functions is reachable is equal to

$$1 - \sum_{1 \le i < N, \, n_i \ne 1} (N - i) \cdot t + O(t^2) =$$

$$= 1 - \left(\frac{N(N-1)}{2} - \sum_{1 \le i < N, \, n_i = 1} (N - i) \right) \cdot t + O(t^2).$$

Proof.
Since the transfer functions of the systems are strictly proper, one has $\deg(P_i) < \deg(Q_i) = n_i$ for $i = 1, \dots, N$. This effects that if $n_i = 1$, P_i and Q_j are not coprime if and only if $P_i \equiv 0$, which is a criterion that is independent of Q_j and has probability t. Therefore, if one proceeds as in the proof of Theorem 32, one could remove all $N - i$ edges terminating at P_i from the graph Γ, which we considered when determining A_{N+1}, and in return add $\tilde{N}_1 \cdot t$ to the hereby obtained probability where $\tilde{N}_1 := |l \in \{1, \dots, N - 1\} \mid n_l = 1|$. Hence, $\Pr(A_{N+1}) = \sum_{1 \le i < N, \, n_i \ne 1}(N - i) \cdot t + \tilde{N}_1 \cdot t + O(t^2)$. Moreover, in comparison with Theorem 32 and its proof, the asymptotic expressions for $A_{i,N+1}$ with $i \le N$ and $n_i = 1$ are changed. It holds $A_{i,N+1} = \{P_i \equiv 0\}$ and hence, $\Pr(A_{i,N+1}) = t$. Because the number of sets $A_{i,N+1}$ with this property is equal to $\tilde{N}_1$, one gets $\sum_{1 \le i < N, \, n_i \ne 1}(N-i) \cdot t + O(t^2)$ for the complementary probability. $\qquad\square$

Remark 20.
The plausibility of the preceeding formula could be seen as follows: If $n_i = 1$, the coprimeness of P_i and Q_i implies that P_i is a nonzero constant. Therefore, it is not possible that P_i and Q_j with $j > i$ are not coprime and one has the corresponding number of possibilities less for $\mathcal{R}_{N+1}$ to be not left prime.

3.3.2 Series Connection of two Systems

In this subsection, a series connection of two arbitrary systems should be considered, which is given by the following equations:

$$
\begin{aligned}
x_1(\tau + 1) &= A_1 x_1(\tau) + B_1 u(\tau) \\
x_2(\tau + 1) &= A_2 x_2(\tau) + B_2 C_1 x_1(\tau) + B_2 D_1 u(\tau)
\end{aligned}
\tag{3.6}
$$

The following theorem provides an asymptotic estimation for the probability of reachability for such an interconnection.

Theorem 34.
The probability that the series connection of two minimal systems $(A_i, B_i, C_i, D_i) \in \mathbb{F}^{n_i \times n_i} \times \mathbb{F}^{n_i \times m_i} \times \mathbb{F}^{p_i \times n_i} \times \mathbb{F}^{p_i \times m_i}$ for $i = 1, 2$ with $p_1 = m_2$ is reachable is

$$
1 - t^{m_1} + O(t^{m_1+1}).
$$

Proof.
Using (3.5), one has to compute

$$
1 - \frac{\Pr(A_3) - \Pr(A_{13}) - \Pr(A_{23}) + \Pr(A_{123})}{1 - \Pr(A_1) - \Pr(A_2) + \Pr(A_{12})}.
$$

From Theorem 12, one knows that $\Pr(A_i) = t^{p_i} + O(t^{p_i+1}) = O(t)$ for $i = 1, 2$. To compute $\Pr(A_3)$, one considers $G := [P_1 \ Q_2] \begin{bmatrix} I & 0 \\ 0 & U \end{bmatrix} = [P_1 \ Q_2 U] = [P_1 \ Q_2^H]$, where U is the unimodular matrix such that Q_2^H is in Hermite form. This transformation does not change left primeness because $\begin{bmatrix} I & 0 \\ 0 & U \end{bmatrix}$ is unimodular, too. Let $z_0 \in \overline{\mathbb{F}}$ such that $G(z_0)$ is singular. Applying the method of iterated row operations (see Lemma 9 (b)), one achieves that there exist $k \in \{1, \dots, p_1\}$, a set of column indices $\{j_1, \dots, j_{k-1}\} \subset \{1, \dots, m_1 + p_1\}$ and values $\lambda_r \in \mathbb{F}(z_0)$, which (only) depend on entries g_{ij} of G with $j \in \{j_1, \dots, j_{k-1}\}$ and on z_0, such that

$$
g_{kj}(z_0) = \sum_{r=1}^{k-1} g_{rj}(z_0) \cdot \lambda_r \qquad \text{for } j \in \{1, \dots, m_1 + p_1\} \setminus \{j_1, \dots, j_{k-1}\}.
\tag{3.7}
$$

Similiar to the proof of Theorem 19, one could conclude $g_{ij} \equiv 0$ for $i + m_1 < j \leq m_1 + p_1$ as Q_2^H is lower triangular, and $\{j_1, \dots, j_{k-1}\} \subset \{1, \dots, m_1 + k - 1\}$. Hence, one has the conditions $(Q_2^H)_{kk}(z_0) = 0$, which implies $\kappa_{m_2-k+1}^{(2)} \geq 1$, and

$$g_{kj}(z_0) = \sum_{r=1}^{k-1} g_{rj}(z_0) \cdot \lambda_r \quad \text{for } j \in \{1,\ldots,m_1+k-1\} \setminus \{j_1,\ldots,j_{k-1}\}. \quad (3.8)$$

Thus, the m_1 polynomials g_{kj} with $j \in \{1,\ldots,m_1+k-1\} \setminus \{j_1,\ldots,j_{k-1}\}$ are not fixed to zero due to degree restrictions but are fixed at z_0 by the remaining polynomials of G. We fix $g := g_{z_0}$ and apply Lemma 7 (a) and (b) with $\tilde{w}_j \equiv 1$ and $w_j := \sum_{r=1}^{k-1} g_{rj} \cdot \lambda_r$ for $j \in \{1,\ldots,m_1+k-1\} \setminus \{j_1,\ldots,j_{k-1}\}$. One obtains a probability that is $O(\varphi_g \cdot t^{g+m_1}) = O(t^{m_1})$ for the above conditions. Consequently, the probability is $O(t^{m_1+1})$ if one has no simple form. If Q_2^H is in simple form, one gets the condition

$$m_1 > \mathrm{rk} \begin{bmatrix} p_{11}^{(1)} & \cdots & p_{1,m_1}^{(1)} & I_{p_1-1} & & 0 \\ \vdots & & \vdots & & & \\ p_{p_1,1}^{(1)} & \cdots & p_{p_1,m_1}^{(1)} & q_{p_1,1}^{H_2} & \cdots & q_{p_1,p_1}^{H_2} \end{bmatrix} (z_0) =$$

$$= p_1 - 1 + \mathrm{rk} \begin{bmatrix} p_{p_1,1}^{(1)} - \sum_{i=1}^{p_1-1} p_{i1}^{(1)} q_{p_1,i}^{H_2} & \cdots & p_{p_1,m_1}^{(1)} - \sum_{i=1}^{p_1-1} p_{i,m_1}^{(1)} q_{p_1,i}^{H_2} & q_{p_1,p_1}^{H_2} \end{bmatrix} (z_0),$$

which is equivalent to

$$p_{p_1,1}^{(1)}(z_0) - \sum_{i=1}^{p_1-1} p_{i1}^{(1)} q_{p_1,i}^{H_2}(z_0) = \ldots = p_{p_1,m_1}^{(1)}(z_0) - \sum_{i=1}^{p_1-1} p_{i,m_1}^{(1)} q_{p_1,i}^{H_2}(z_0) = q_{p_1,p_1}^{H_2}(z_0) = 0.$$

Now, one proceeds as in the part of the proof for Theorem 12 concerning simple form, with $u \equiv 1$ and $s^{(j)} = \sum_{i=1}^{p_1-1} p_{i,j}^{(1)} q_{p_1,i}^{H_2}$ for $j = 1,\ldots,m_1$. Doing this, one gets that the probability that $G(z_0)$ is singular is upper bounded by $g \cdot \varphi_g \cdot t^g \cdot \prod_{j=1}^{m_1} t^{\min(g,\nu_j^{(1)}+1)}$. As $\sum_{j=1}^{m_1} \nu_j^{(1)} = n_1 \geq 1$, there exists $j \in \{1,\ldots,m_1\}$ with $\nu_j^{(1)} \geq 1$ and therefore, the probability is $O(t^{m_1+1})$ for $g \geq 2$. Again as in the proof of Theorem 12, one gets that this probability is $t^{m_1} + O(t^{m_1+1})$ for $g = 1$ and hence, $\Pr(A_3) = t^{m_1} + O(t^{m_1+1})$.

It remains to show that $\Pr(A_{13}) = O(t^{m_1+1}) = \Pr(A_{23})$. For this, one has to consider only simple form of Q_1^H and Q_2^H since one knows that the probability of being not in simple from and lying in A_3 is already $O(t^{m_1+1})$.

For A_{13}, this leads to the condition that there exist $z_0, \tilde{z}_0 \in \overline{\mathbb{F}}$ such that

$$q_{m_1,m_1}^{H_1}(z_0) = \sum_{l=1}^{m_1} p_{1l}^{(1)} u_{l,m_1}^{(1)}(z_0) = \cdots = \sum_{l=1}^{m_1} p_{p_1,l}^{(1)} u_{l,m_1}^{(1)}(z_0) = 0 \quad (3.9)$$

$$p_{p_1,1}^{(1)}(\tilde{z}_0) - \sum_{i=1}^{p_1-1} p_{i1}^{(1)} q_{p_1,i}^{H_2}(\tilde{z}_0) = \cdots = p_{p_1,m_1}^{(1)}(\tilde{z}_0) - \sum_{i=1}^{p_1-1} p_{i,m_1}^{(1)} q_{p_1,i}^{H_2}(\tilde{z}_0) =$$

$$= q_{p_1,p_1}^{H_2}(\tilde{z}_0) = 0. \quad (3.10)$$

One could assume $z_0, \tilde{z}_0 \in \mathbb{F}$. This is possible since if $\tilde{z}_0 \notin \mathbb{F}$, one has a probability of $O(t^{m_1+1})$ by only using the conditions containing $\tilde{z}_0$ and if $z_0 \notin \mathbb{F}$, one has a factor that is $O(t^{p_1+1}) = O(t^2)$ from the conditions containing z_0 and a factor that is $O(t^{m_1-1})$ from the conditions containing $\tilde{z}_0$ without the condition on $p_{p_1,l_0}^{(1)}$, where l_0 is chosen such that $u_{l_0,m_1}^{(1)}(z_0) \neq 0$.

For $z_0 = \tilde{z}_0$, one has a probability of $O(t^{m_1})$ for (3.10) and (amongst others) the additional factor t because of $q_{m_1,m_1}^{H_1}(z_0) = 0$. Thus, one obtains a probability that is $O(t^{m_1+1})$.

For $z_0 \neq \tilde{z}_0$, $p_{p_1,l_0}^{(1)}$ is fixed at two different values. If $\nu_{l_0}^{(1)} \geq 1$, i.e. $q_{l_0,l_0}^{(1)} \neq 1$, one could apply Lemma 7 (c) and gets the factor t^2 for the number of possibilities of $p_{p_1,l_0}^{(1)}$. This is possible since if $q_{l_0,l_0}^{(1)} \equiv 1$, all other entries of Q_1 in row l_0 would be zero, due to the Kronecker-Hermite form of Q_1. Thus, the l_0-th row of $Q_1^H = Q_1 U_1$ would be equal to the l_0-th row of U_1. But (3.9) implies that the last column of $Q_1^H(z_0)$ is equal to zero, which leads to $u_{l_0,m_1}^{(1)}(z_0) = 0$. However, this is a contradiction to the choice of l_0. In summary, there are $t^{-1}(t^{-1} - 1)$ possibilities for the choice of $z_0, \tilde{z}_0 \in \mathbb{F}$, the factor t^2 for fixing $p_{p_1,l_0}^{(1)}$ and the factor $t^{m_1+p_1}$ for fixing the other polynomials, which leads a probability that is $O(t^{m_1+p_1}) = O(t^{m_1+1})$.

For A_{23}, one has the condition that there exist $z_0, \tilde{z}_0 \in \mathbb{F}$ such that

$$q_{p_1,p_1}^{H_2}(z_0) = \sum_{l=1}^{p_1} p_{1l}^{(2)} u_{l,p_1}^{(2)}(z_0) = \cdots = \sum_{l=1}^{p_1} p_{p_2,l}^{(2)} u_{l,p_1}^{(2)}(z_0) = 0$$

$$p_{p_1,1}^{(1)}(\tilde{z}_0) - \sum_{i=1}^{p_1-1} p_{i1}^{(1)} q_{p_1,i}^{H_2}(\tilde{z}_0) = \cdots = p_{p_1,m_1}^{(1)}(\tilde{z}_0) - \sum_{i=1}^{p_1-1} p_{i,m_1}^{(1)} q_{p_1,i}^{H_2}(\tilde{z}_0) =$$

$$= q_{p_1,p_1}^{H_2}(\tilde{z}_0) = 0.$$

One could argue analogous to the consideration of A_{13}. The only difference is that now, the polynomial $q_{p_1,p_1}^{H_2}$ is fixed to zero at two values. For $z_0 = \tilde{z}_0$, one has $p_1 + 1 + m_1$ fixed polynomials and hence, a probability that is $O(t^{m_1+1})$. If $z_0 \neq \tilde{z}_0$ and $\deg(q_{p_1,p_1}^{H_2}) = 1$, one has a contradiction because a linear polynomial cannot have two distinct zeros; thus, this case cannot occur and therefore, its probability is equal to zero. If $z_0 \neq \tilde{z}_0$ and $\deg(q_{p_1,p_1}^{H_2}) \geq 2$, one could argue as for the calculation of $\Pr(A_{13})$ and gets that $\Pr(A_{23})$ is $O(t^{m_1+p_2}) = O(t^{m_1+1})$.

Putting all achieved results together, one obtains that the overall probability is equal to

$$1 - \frac{t^{m_1} + O(t^{m_1+1})}{1 - O(t)} = 1 - t^{m_1} + O(t^{m_1+1}).$$

$\square$

Corollary 14.
The probability that a series connection of two minimal systems is observable is equal to $1 - t^{p_2} + O(t^{p_2+1})$.

Proof.
Using the criterion on $\hat{Q}$ and $\hat{P}$ from Example 1 (b), one needs to compute the probability that $\hat{Q}_1$ and $\hat{P}_2$ are right coprime if $\hat{Q}_i$ and $\hat{P}_i$ are left coprime for $i = 1, 2$. One could assume that $\hat{Q}_i^\top$ is in Kronecker-Hermite form and that the column degrees of $\hat{P}_i^\top$ are less or equal to the column degrees of $\hat{Q}_i^\top$ for $i = 1, 2$. Moreover, two matrices are right/left coprime if and only if the transposed matrices are left/right coprime. Thus, to compute the probability of observability with the help of the criterion on $\hat{Q}$ and $\hat{P}$ is equivalent to computing the probability of reachability with the help of the criterion on Q and P with $Q_i = \hat{Q}_j^\top$ and $P_i = \hat{P}_j^\top$ for $1 \leq i \neq j \leq 2$. Hence, one has to replace m_1 by p_2 in the formula for reachability of the series connection. $\qquad\square$

As in the preceding subsection, which dealt with series connections of SISO systems, the series connection of two minimal systems with strictly proper transfer functions, i.e. with $D_1 = D_2 = 0$, has a different probability of reachability than in the proper case. In the following theorems, we will investigate the strictly proper case, where it will turn out that the asymptotic probability does not only depend on m_1 but also on the sizes of the other involved system matrices.

Theorem 35.
For $m_1 = n_1 = 1$, the probability that a series connection of two minimal systems with strictly proper transfer functions is reachable is equal to 1 if $p_1 = 1$ and equal to $1 - t + O(t^2)$ if $p_1 \geq 2$.

Proof.
Here, Q_1 is scalar and P_1 a vector consisting of p_1 constants. Thus, if $p_1 = 1$, i.e. P_1 and Q_2 are scalar, too, P_1 and Q_2 are left coprime if and only if P_1 and Q_1 are right coprime, namely if and only if $P_1 \neq 0$. Consequently, the minimality of the node systems implies the reachability of the series connected system and hence, the considered probability is equal to 1.
Now, examine the case $p_1 \geq 2$. Since Q_1 is a linear polynomial, no entry of P_1 is fixed to zero due to degree conditions. Therefore, $\Pr(A_3) = t^{m_1} + O(q^{m_1+1}) = t + O(t^2)$ as in the proper case. Since P_1 is a vector consisting of p_1 constants, Q_1 and P_1 are right coprime if and only if P_1 is not the zero vector. But if P_1 is the zero vector, it is not left prime with Q_2. Consequently, $\Pr(A_{13}) = \Pr(A_1) = t^{p_1}$ (considering the proof of Theorem 34 this means that (3.9) implies (3.10) since fixing a constant to zero at one point effects that it is zero at every point). Thus, the equation $\Pr(A_{13}) = O(t^{m_1+1}) = O(t^2)$, which one had in the proper case, is only valid for $p_1 \geq 2$. From Remark 8, one knows $\Pr(A_2) = t^{p_2}$ and with the same argumentation as in the proper case, one has $\Pr(A_{23}) = O(t^2)$. In summary, the overall probability is equal to

$$1 - \frac{t + O(t^2)}{1 + O(t)} = 1 - t + O(t^2).$$

$\qquad\square$

Theorem 36.
For $m_1 = 1$, $n_1 \geq 2$, the probability for a series connection of two minimal systems with strictly proper transfer functions to be reachable is equal to
$1 - t + O(t^2)$.

Proof.
This theorem could be proven analogous to Theorem 34 since none of the entries of P_1 has to be a constant due to degree conditions. □

Theorem 37.
With the same notations as above but $D_1 = D_2 = 0$, one has for $m_1 \geq 2$:

$$\Pr(A_3) = t^{m_1} + t^{n_1} + O(t^{\min(m_1,n_1)+1})$$
$$\Pr(A_i) = t^{p_i} + O(t^{(p_i+1)}) \quad \text{for } i = 1, 2$$
$$\Pr(A_{23}) = O(t^{\min(m_1,n_1)+1}).$$

Proof.
The second statement of the theorem follows from Remark 8.
Now, consider $\Pr(A_3)$. Since $P_1 Q_1^{-1}$ is strictly proper, every column i of P_1, for which $q_{ii}^{(1)} \equiv 1$, consists only of fixed zeros. Hence, these columns could be deleted considering the right primeness of $[P_1 \; Q_2]$. With the remaining matrix, one could proceed as in the proper case. Thus, if N_1 is the number of deleted columns, one gets $\Pr(A_3) = t^{m_1-N_1} + O(t^{m_1-N_1+1})$.
Consequently, one has to count or at least to estimate the number of Kronecker-Hermite forms with fixed column degrees $\nu_1, \ldots, \nu_{m_1}$, of which N_1 are equal to zero. If these parameters are fixed, the number of Kronecker-Hermite forms is a power of t^{-1}. We call the corresponding exponent the dimension of the set of matrices with these column degrees. Since we are only interested in the leading term of $\Pr(A_3)$ and the order of the values for the column degrees does not change the number of column degress that are equal to zero, one could assume $\nu_1 \leq \ldots \leq \nu_{m_1}$. This is possible because in column i the entries above the diagonal have strictly lower degree then ν_i but the polynomials beyond the diagonal could have degree equal or less to ν_i (see the definition of the Kronecker-Hermite form). This effects that the dimension is largest for $\nu_1 \leq \ldots \leq \nu_{m_1}$ among all permutations of the values for the column degrees. Moreover, for every other order of $\nu_1, \ldots, \nu_{m_1}$, the dimension is as much smaller as the number of permutations of degrees of adjacent columns that is needed to achieve increasing order. To see this, assume $\nu_i > \nu_{i+1}$. If one exchanges these two indices, the number of total possibilities for all entries but $q_{i,i+1}^{(1)}$ and $q_{i+1,i}^{(1)}$ is not effected (there are just entries in this set which exchange their number of possibilities with other entries from that set). Futhermore, one has $\deg(q_{i,i+1}^{(1)}) \leq \min(\nu_i - 1, \nu_{i+1} - 1) = \nu_{i+1} - 1$ and $\deg(q_{i+1,i}^{(1)}) \leq \min(\nu_{i+1} - 1, \nu_i) = \nu_{i+1} - 1$ before the exchange and $\deg(q_{i,i+1}^{(1)}) \leq \min(\nu_{i+1} - 1, \nu_i - 1) = \nu_{i+1} - 1$ and $\deg(q_{i+1,i}^{(1)}) \leq \min(\nu_i - 1, \nu_{i+1}) = \nu_{i+1}$ afterwards. Hence, the dimension has increased by one.

To count the number of Kronecker-Hermite forms with increasing column degrees of which N_1 are zero and whose sum is equal to n_1, one starts with $\nu_1 = \ldots = \nu_{m_1-1} = 0$, $\nu_{m_1} = 1$ and successively increases one of the values for the ν_i until their sum reaches n_1 and where in each step, the values for the column degrees remain increasing, i.e. if one value for the column degrees occurs several times, one could only increase the last one of them. Doing this, we want to determine the effect on the dimension of increasing ν_i from w to $w + 1$. First, consider what happens in row i. For $j > i$, i.e. $\nu_j \geq w + 1$, the upper bound for the degree of the (i, j)-entry increases from $w - 1$ to w. For $j < i$, i.e. $\nu_j \leq w$, this increasing is only possible if $\nu_j = w$. Now, look at column i. For $j > i$, a increasing of the degree of the (j, i)-entry from w to $w + 1$ is possible if and only if $\nu_j > w + 1$. For $j < i$, an increase from $w - 1$ to w is not possible since it would require $\nu_j > w$, a contradiction to the increasing order of the column degrees. In summary, the dimension is increased by $\underbrace{m_1 - 1 - |j \in \{1, \ldots, m_1\} \mid \nu_j < w|}_{\text{effect in row } i} + \underbrace{|j \in \{1, \ldots, m_1\} \mid \nu_j > w + 1|}_{\text{effect in column } i}$. It could be seen easily that this term decreases with w and therefore, is largest for $w = 0$, namely $m_1 - 1 + |j \in \{1, \ldots, m_1\} \mid \nu_j > 1|$. But in this case, after the increasing of a column degree from 0 to 1, one has a column of fixed zeros (due to degree restrictions) less in $[P_1 \ Q_2]$. This decreases the asymptotic probability $\Pr(A_3) = t^{m_1-N_1} + O(t^{m_1-N_1+1})$ by the factor t. For $w \neq 0$, this probability remains unchanged.

If we start at $\nu_1 = \ldots = \nu_{m_1-1} = 0$, $\nu_{m_1} = 1$, it is only possible to choose $w = 1$ and increase ν_{m_1} or to choose $w = 0$ and increase ν_{m_1-1}. In both cases, the term for the effect on the column is equal to zero. As long as there are at least two column degrees equal to 0, choosing $w = 1$ increases the dimension by at least 2 less than choosing $w = 0$. Hence, since the difference in the probability $\Pr(A_3)$ is only of dimension one, it is sufficient to consider the case that the column degree with the largest index among those that are equal to zero is increased as long as there is still a column degree equal to zero after this increasing.

Consequently, for $n_1 < m_1$, the only relevant Kronecker-Hermite forms are those with $\nu_1 = \ldots = \nu_{m_1-n_1} = 0$ and $\nu_{m_1-n_1+1} = \ldots = \nu_{m_1} = 1$. In this case, one has $N_1 = m_1 - n_1$ and therefore, $\Pr(A_3) = t^{n_1} + O(t^{n_1+1})$.

For $n_1 = m_1$, one could choose $\nu_1 = 0$, $\nu_2 = \ldots = \nu_{m_1} = 1$, as starting point for increasing column degrees. Doing this, choosing $w = 0$ increases the dimension by $m_1 - 1$, while choosing $w = 1$ increases the dimension by $m_1 - 2$. This difference compensates the difference in the estimation for $\Pr(A_3)$. Hence, one has to consider both $\nu_1 = \ldots = \nu_{m_1} = 1$ and $\nu_1 = 0$, $\nu_2 = \ldots = \nu_{m_1-1} = 1$, $\nu_{m_1} = 2$. In the first case, $\Pr(A_3) = t^{m_1} + O(t^{m_1+1})$, in the second case, $\Pr(A_3) = t^{m_1-1} + O(t^{m_1})$. Since one has a factor t in the second case, due to the fact that the dimension is 1 less, the overall probability is equal to $\Pr(A_3) = 2t^{m_1} + O(t^{m_1+1})$.

For $n_1 > m_1$, one has to consider forms obtained by increasing degrees starting with $\nu_1 = \ldots = \nu_{m_1} = 1$ or starting with $\nu_1 = 0$, $\nu_2 = \ldots = \nu_{m_1-1} = 1$, $\nu_{m_1} = 2$. In the first case, one only has the possibility $w = 1$. In the second case, one has a increase of the dimension by m_1 for $w = 0$, by $m_1 - 2$ for $w = 1$ and no increase for $w = 2$. Thus, the cases but $w = 0$ have a dimension that is at least by 2 smaller than the dimension for $w = 0$ and since the difference in $\Pr(A_3)$ is only of dimension one, they could be neglected. Hence, in either case, one gets $\nu_1 = \ldots = \nu_{m_1-1} = 1$, $\nu_{m_1} = 2$. No matter in which way one continues, only Kronecker-Hermite forms with no fixed ones on the diagonal matter. Consequently, $\Pr(A_3) = t^{m_1} + O(t^{m_1+1})$ as in the proper case. The identity $\Pr(A_{23}) = O(t \cdot \Pr(A_3))$ is valid with the same argumentation as in the proper case. $\qquad \square$

The difficulty with the determination of the probability of reachability for a series connection of strictly proper systems lies in the determination of $\Pr(A_{13})$. What we have computed so far, shows that it does not matter if D_2 is equal to zero or not. But this is obvious since reachability of a system does not depend on the output. D_1 matters because it influences the input of the second system.

Theorem 38.
For $m_1, p_1 \geq 2$, the probability that a series connection of two minimal systems with strictly proper transfer functions is reachable is equal to

$$1 - t^{m_1} - t^{-n_1} + O(t^{\min(m_1, n_1)+1}).$$

Proof.
For $p_1 \geq 2$, it is easy to compute $\Pr(A_{13})$. One deletes the zero columns of P_1 and then proceeds as in the proof of Theorem 34. From equation (3.9) one knows that $p_1 + 1$ polynomials are fixed at z_0. According to (3.10), $m_1 - N_1 + 1$ polynomials are fixed at $\tilde{z}_0$, from which $m_1 - N_1$ do not belong to the fixed polynomials from (3.9). Thus, it follows from previous computations that for fixed column degrees, $\Pr(A_{13}) = O(t^{p_1+m_1-N_1-1}) = O(t^{m_1-N_1+1}) = O(t \cdot \Pr(A_3))$ for $p_1 \geq 2$; see end of the first paragraph of the previous proof. Hence, the statement follows from the results of the preceding theorem. $\qquad \square$

Theorem 39.
For $m_1 \geq 2$, $p_1 = 1$, the probability that the series connection of two minimal systems with strictly proper transfer functions is reachable is equal to

$$\begin{array}{ll} 1 & \text{for } n_1 = 1 \\ 1 - t^{m_1} + O(t^{m_1+1}) & \text{for } n_1 \geq 2. \end{array}$$

Proof.
It remains to compute $\Pr(A_{13})$ in this case.
Here, P_1 is a row vector of length m_1 and Q_2 is scalar. If Q_1 is in Kronecker-Hermite form with $\nu_i \in \{0, 1\}$ for $i \in \{1, \ldots, m_1\}$, P_1 only consists of constants. For Q_1 and P_1 to be right coprime, it is necessary that $P_1 \neq 0$. This condition, however, is sufficient for P_1 and Q_2 to be left coprime. Hence, the probability of reachability for the interconnected system is equal to 1; in particular, this probability is equal to 1 if $n_1 = 1$ because in this case, it always holds $\nu_i \in \{0, 1\}$ for $i \in \{1, \ldots, m_1\}$.
Now consider the case that there exists $i \in \{1, \ldots, m_1\}$ with $\nu_i \geq 2$. For A_3, one has the condition that there exists $\tilde{z}_0 \in \bar{\mathbb{F}}$ with $p_1^{(1)}(\tilde{z}_0) = \cdots = p_{m_1}^{(1)}(\tilde{z}_0) = Q_2(\tilde{z}_0) = 0$. We already know that this probability is $O(t^{m_1})$ for $\tilde{z}_0 \in \mathbb{F}$ and $O(t^{m_1+1})$, otherwise.

Therefore, it is sufficient to consider $\tilde{z}_0 \in \mathbb{F}$ and the case that Q_1^H is of simple form. For A_1, without restriction, assume that Q_1 has no column degree equal to zero; see beginning of the proof of Theorem 12. Moreover, one could assume $\nu_1 \leq \ldots \leq \nu_{m_1}$; see proof of Theorem 37. Proceeding as in the proof of Theorem 12, one gets the condition that there exists $z_0 \in \overline{\mathbb{F}}$ with $q_{m_1,m_1}^H(z_0) = \sum_{l=1}^{m_1} p_l^{(1)} u_{l,m_1}^{(1)}(z_0) = 0$. Clearly, the subset of A_{13} with $z_0 = \tilde{z}_0$ has a probability that is $O(t^{m_1+1})$. Thus, it remains to consider $z_0 \neq \tilde{z}_0$. Let $d \leq m_1 - 1$ be the number of $i \in \{1, \ldots, m_1\}$ with $\nu_i < 2$. Then, A_3 implies $p_1 \equiv \ldots \equiv p_d \equiv 0$. If $p_{d+1}^{(1)}(z_0) = \cdots = p_{m_1}^{(1)}(z_0) = 0$, the polynomial $p_{m_1}^{(1)}$, which degree bound is equal to $\nu_{m_1} \geq 2$ per assumption, is fixed at two values and therefore, $\Pr(A_{13}) = O(t \cdot \Pr(A_3)) = O(t^{m_1+1})$. If there exists $i \in \{d+1, \ldots, m_1\}$ such that $p_i^{(1)}(z_0) \neq 0$, consider the matrix $G := \begin{pmatrix} Q_1 \\ P_1 \end{pmatrix}$ and nullify column i of $G(z_0)$ by adding appropriate multiples of the last row to the other rows of this matrix. Afterwards delete the last row and i-th column of this matrix and denote the resulting matrix by $\tilde{G}(z_0)$. Obviously, $\tilde{G}(z_0)$ is singular if and only if $G(z_0)$ is singular. Now, one applies the method of iterated column operations (see Lemma 9 (a)) to the matrix $\tilde{G}(z_0)$ and gets that at least $k \geq 2$ polynomials from $\tilde{G}(z_0)$ are fixed by the values of the other entries of this matrix. Since we deleted the last row of $G(z_0)$ and P_1 consists only of one row, these conditions have the form $q_j^{(1)}(z_0) \cdot h_j(z_0) - f_j(z_0) = 0$ for $j = 1, \ldots, k$ where the polynomials $q_j^{(1)}$ are entries of Q_1 (that are not fixed constants due to degree conditions because we assumed that no column degree is equal to zero) and h_j and f_j are products of the other entries of G with $h_j(z_0) \neq 0$. Hence, either $q_j^{(1)} \cdot h_j - f_j \equiv 0$ or there exists an upper bound for g_{z_0} determined by the column degrees of Q_1. In the first case, the polynomials $q_j^{(1)}$ for $j = 1, \ldots, k$ are completely determined by the other entries of G and therefore, one even has $\Pr(A_{13}) = O(t^2 \cdot \Pr(A_3))$ because $\Pr(A_3)$ is independent of Q_1. In the second case, for each value of g_{z_0} (for which there are only finitely many possibilities), one applies Lemma 7 (b) to the conditions $q_j^{(1)}(z_0) \cdot h_j(z_0) - f_j(z_0) = 0$ for $j = 1, \ldots, k$, which gives a factor for the possibilities for Q_1 that is $O(t^{k-1}) = O(t)$. Again since entries of Q_1 do not matter for $\Pr(A_3)$, one has $\Pr(A_{13}) = O(t \cdot \Pr(A_3)) = O(t^{m_1+1})$.

For $n_1 > m_1$, there always exists $i \in \{1, \ldots, m_1\}$ with $\nu_i \geq 2$ and therefore, one could conclude that the probability of reachability for the interconnected system is equal to $1 - t^{m_1} - t^{n_1} + O(t^{(\min(m_1,n_1)+1)}) = 1 - t^{m_1} + O(t^{m_1+1})$ if one uses the results from Theorem 37.

If $1 < n_1 \leq m_1$ and $\nu_i \in \{0, 1\}$ for $i = 1, \ldots, m_1$, A_3 is equivalent to $P_1 \equiv 0$, which implies A_1. Hence, $\Pr(A_{13}) = \Pr(A_3)$ and $\Pr(A_{23}) = \Pr(A_{123})$. Thus, the probability that the interconnected system is not reachable is zero; see also beginning of the whole proof. If there exists $i \in \{1, \ldots, m_1\}$ with $\nu_i \geq 2$, we saw in the preceding paragraph that the leading coefficient of this probability is equal to the leading coefficient of $\Pr(A_3)$. Therefore, considering different structures of Kronecker-Hermite forms, one proceeds similarly to the proof of Theorem 37 but starts with $\nu_1 = \ldots = \nu_{m_1-1} = 0$, $\nu_{m_1} = 2$. With the same argumentation as in that proof, one could show that the only structure that matters for the leading term of the probability that the interconnected system is not reachable is $\nu_1 = \ldots = \nu_{m_1-n_1+1} = 0$, $\nu_{m_1-n_1+2} = \ldots = \nu_{m_1-1} = 1$, $\nu_{m_1} = 2$. However, the structure with largest dimension amongst all Kronecker-Hermite forms with these parameters n_1 and m_1 has $\nu_1 = \ldots = \nu_{m_1-n_1} = 0$, $\nu_{m_1-n_1+1} = \ldots = \nu_{m_1} = 1$ as values for its column degrees. Hence, computing the loss in the dimension, one has to look at the effect of choosing $w = 1$ instead of $w = 0$ (see proof of Theorem 37), when increasing a column degree starting from $\nu_1 = \ldots = \nu_{m_1-n_1+1} = 0$, $\nu_{m_1-n_1+2} = \ldots = \nu_{m_1} = 1$. Using the formula from the proof of Theorem 37, one gets that this difference in the dimension is equal to $|j \in \{1, \ldots, m_1\} \mid \nu_j < 1| = m_1 - n_1 + 1$. Now, distinguish two subcases.

For $n_1 = m_1$, we already saw in the proof of Theorem 37 that this loss of 1 in the dimension is compensated by a higher value for $\Pr(A_3)$ due to an additional column of zeros in P_1 (which is here just an additional zero entry because of $p_1 = 1$). But since one has $\Pr(A_{13}) = \Pr(A_3)$ and $\Pr(A_{23}) = \Pr(A_{123})$ for the structure $\nu_1 = \ldots = \nu_{m_1} = 1$, the leading term of the probabiliy that the interconnection is not reachable is t^{m_1} (and not $2t^{m_1}$, which is the leading term of $\Pr(A_3)$). Thus, using the results of Theorem 37, one gets an overall probability of $1 - t^{m_1} + O(t^{m_1+1})$.

For $n_1 < m_1$, one has to multiply the term for $\Pr(A_3)$ from Theorem 37 by the factor $t^{m_1-n_1+1}$ for the decreased dimension and afterwards, by the factor t^{-1} due to the additional zero entry in P_1 compared with the structure of largest dimension. Consequently, one gets $t^{n_1+m_1-n_1+1-1}$ for the leading term of the probability that the interconnected system is not reachable and therefore, $1 - t^{m_1} + O(t^{m_1+1})$ for the probability of reachability. $\qquad\square$

Summarizing the results of the preceding theorems, on gets the following theorem:

Theorem 40.
The probability that the series connection of two minimal systems with strictly proper transfer functions is reachable is equal to

$$
\begin{array}{ll}
1 & \text{for } p_1 = n_1 = 1, \\
1 - t^{m_1} + O(t^{m_1+1}) & \text{for } p_1 = 1, \ n_1 > 1 \text{ or } p_1 \geq 2, \ n_1 > m_1 \\
& \text{or } p_1 \geq 2, \ m_1 = 1, \\
1 - 2t^{m_1} + O(t^{m_1+1}) & \text{for } p_1 \geq 2, \ n_1 = m_1 \neq 1, \\
1 - t^{n_1} + O(t^{n_1+1}) & \text{for } p_1 \geq 2, \ n_1 < m_1.
\end{array}
$$

3.4 Circular Interconnection

As we saw in Example 1 (c), for circular interconnection, one has the same coprimeness condition as for series connection, namely that

$$
\mathcal{R}_N := \begin{bmatrix}
P_1 & Q_2 & 0 & \cdots & 0 \\
0 & P_2 & Q_3 & \ddots & \vdots \\
\vdots & \ddots & \ddots & \ddots & 0 \\
0 & \cdots & 0 & P_{N-1} & Q_N
\end{bmatrix}
$$

has to be left prime. However, one has to assume $\det(I - D_N \cdots D_1) \neq 0$ to be able to apply Theorem 10. Therefore, one has to investigate if this condition influences the probability of reachability. Doing this, we will obtain that the probability stays asymptotically the same. As in the previous section, for $i = 1, \ldots, N$, define $A_i := \{(P_i, Q_i) \text{ not right coprime}\}$, $A_{N+1} := \{\mathcal{R}_N \text{ not left prime}\}$ and $A_{ij} := A_i \cap A_j$. But here, one additionally needs $A_0 := \{\det(I - D_N \cdots D_1) = 0\}$. Consequently, one gets that the probability of reachability is equal to

$$\frac{1 - \Pr(\cup_{i=0}^{N+1} A_i)}{1 - \Pr(\cup_{i=0}^{N} A_i)} = \frac{1 - \sum_{\emptyset \neq I \subset \{0,\ldots,N+1\}} (-1)^{|I|-1} \Pr(A_I)}{1 - \sum_{\emptyset \neq I \subset \{0,\ldots,N\}} (-1)^{|I|-1} \Pr(A_I)} =$$

$$= 1 - \frac{\Pr(A_{N+1}) + O(\sum_{i=0}^{N} \Pr(A_{i,N+1}))}{1 - \Pr(A_0) + O(t)} \tag{3.11}$$

since one already knows from the previous section that $\Pr(A_i) = O(t)$ for $i = 1, \ldots, N$.

3.4.1 Circular Interconnection of SISO Systems

As done for series connection, we start considering a circular interconnection of SISO systems, i.e. P_i and Q_i are scalar for $i = 1, \ldots, N$. Since for this case, we know from the proof for a series connection of SISO systems that $\Pr(A_{N+1}) = \frac{N(N-1)}{2}t + O(t^2)$ as well as $\Pr(A_{i,N+1}) = O(t^2)$ for $i = 1, \ldots, N$, it only remains to estimate $\Pr(A_0)$ and $\Pr(A_{0,N+1})$. Doing this, one gets the following result:

Theorem 41.
The probability that a circular interconnection of N minimal SISO systems is reachable is equal to

$$1 - \frac{N(N-1)}{2} \cdot t + O(t^2).$$

Proof.
It follows from (3.11) and the preceding observations that the considered probability is equal to

$$1 - \frac{\frac{N(N-1)}{2}t + O(t^2) - \Pr(A_{0,N-1})}{1 - \Pr(A_0) + O(t)}.$$

For the computation of $\Pr(A_0)$, one performs the transformation

$\det(I - D_N \cdots D_1) = \lim_{z \to \infty} \left(1 - \frac{P_1 \cdots P_N}{Q_1 \cdots Q_N}(z)\right) = 1 - p_{n_1} \cdots p_{n_N}$, where $p_{n_i} \in \mathbb{F}$ denotes the coefficient of z^{n_i} in $P_i \in \mathbb{F}[z]$. Hence, in the set A_0, p_{n_N} is fixed by the other involved polynomials, which implies $\Pr(A_0) = O(t)$. Moreover, p_{n_N} does not influence $\Pr(A_{N+1})$ and thus, $\Pr(A_{0,N+1}) = \Pr(A_{N+1}) \cdot \Pr(A_0) = O(t^2)$. $\square$

Corollary 15.
The probability that a circular interconnection of N minimal SISO systems is observable is equal to

$$1 - \frac{N(N-1)}{2} \cdot t + O(t^2).$$

3.4.2 Feedback Interconnection

In this subsection, we consider a circular interconnection of two systems with arbitrarily many inputs, i.e. a feedback interconnection.

Theorem 42.
The probability that a feedback interconnection of two minimal systems is reachable is equal to

$$1 - t^m + O(t^{m+1})$$

where $m := m_1 = p_2$ is the number of inputs of the first as well as the number of outputs of the second system. Moreover, set $p := p_1 = m_2$.

Proof.
According to (3.11), the probability of reachability is equal to

$$\frac{1 - \Pr(A_1 \cup A_2 \cup A_3 \cup A_0)}{1 - \Pr(A_1 \cup A_2 \cup A_0)} = 1 - \frac{\Pr(A_3) - \Pr(A_{03}) + O(t \cdot \Pr(A_3))}{1 - \Pr(A_0) + O(t)}$$

since one already knows from the section concerning series connection that $\Pr(A_{13}) = \Pr(A_{23}) = O(t \cdot \Pr(A_3))$. Therefore, it only remains to estimate $\Pr(A_0)$ and $\Pr(A_{03})$. Using $D_i = \lim_{z \to \infty} P_i Q_i^{-1}(z)$ for $i = 1, 2$ as well as the formula

$$\det \begin{bmatrix} Q_2 & P_1 \\ P_2 & Q_1 \end{bmatrix} = \det(Q_2) \cdot \det(I - P_2 Q_2^{-1} P_1 Q_1^{-1}) \cdot \det(Q_1),$$ one gets that condition A_0 is equivalent to the condition

$$\lim_{z \to \infty} \frac{\det \begin{bmatrix} Q_2(z) & P_1(z) \\ P_2(z) & Q_1(z) \end{bmatrix}}{\det(Q_2(z)) \cdot \det(Q_1(z))} = 0.$$

Because Q_1 and Q_2 are in Kronecker-Hermite form with $\deg_j(P_i) \leq \deg_j(Q_i)$ for $j = 1, \ldots, m_i$ and $i = 1, 2$, this is equivalent to the condition that the highest column

degree coefficient matrix $\begin{bmatrix} Q_2 & P_1 \\ P_2 & Q_1 \end{bmatrix}_{hc} = \begin{bmatrix} Q_{c_2} & P_{c_1} & p_{c_1} \\ P_{c_2} & Q_{c_1} & 0 \\ p_{c_2} & q_{c_1} & 1 \end{bmatrix}$ with $Q_{c_2} \in \mathbb{F}^{p \times p}$,

$P_{c_1} \in \mathbb{F}^{p \times (m-1)}$, $P_{c_2} \in \mathbb{F}^{(m-1) \times p}$, $Q_{c_1} \in \mathbb{F}^{(m-1) \times (m-1)}$, $p_{c_1} \in \mathbb{F}^{p \times 1}$, $p_{c_2} \in \mathbb{F}^{1 \times p}$ and $q_{c_1} \in \mathbb{F}^{1 \times (m-1)}$ is not invertible. Note that the Kronecker-Hermite form of Q_i implies that Q_{c_i} is lower triangular with all diagonal elements equal to 1 for $i = 1, 2$.
In the following, it should be shown per induction with respect to $m + p$ that $\Pr(A_0) = O(t)$ and $\Pr(A_{03}) = O(t \cdot \Pr(A_3))$.
The proof for the base clause $m = p = 1$ was already done in the previous subsection. Now, consider the step from $m + p - 1$ to $m + p$.
Assume without restriction that $m \geq 2$. One expands

$\det \begin{bmatrix} Q_2 & P_1 \\ P_2 & Q_1 \end{bmatrix}_{hc} = \begin{bmatrix} Q_{c_2} & P_{c_1} & p_{c_1} \\ P_{c_2} & Q_{c_1} & 0 \\ p_{c_2} & q_{c_1} & 1 \end{bmatrix}$ along the last row and gets $p_{c_2,1} M_1 - \cdots \pm$

$q_{c_1,m-1} M_{p+m-1} \mp M_{p+m}$, where M_i denotes the minor that is formed by all rows but the last and all columns but the i-th. Hence, one could distinguish two cases for A_0. If $M_{p+m} = 0$, one has $A_0 = O(t)$ per induction. If $M_{p+m} \neq 0$, the set $\{(p_{c_2,i}, M_i), \ i = 1, \ldots, p\} \cup \{(q_{c_1,i}, M_{p+i}), \ i = 1, \ldots, m-1\}$ contains an element from $(\mathbb{F} \setminus \{0\})^2$. Hence, there either exists $i \in \{1, \ldots, p\}$ such that $p_{c_2,i}$ is fixed by the other values of the matrix or there exists $i \in \{1, \ldots, m-1\}$ such that $q_{c_1,i}$ is unequal to zero (and hence, in particular, not fixed to zero by degree restrictions) and fixed by the other values of the matrix. Consequently, $A_0 = O(t)$. For $\mathrm{Pr}(A_{03})$, consider the same two cases. For the second case, there is nothing left to do since A_3 is not influenced by the values of the last row of the above matrix and the value that is fixed by condition A_0 is an element of this last row. Thus, A_3 and A_0 are independent and the statement follows. For the first case, i.e $M_{p+m} = 0$, write $P_1 = [\tilde{P}_1 \ p_1]$ with $\tilde{P}_1 \in \mathbb{F}[z]^{p \times (m-1)}$ and $p_1 \in \mathbb{F}[z]^{p \times 1}$. Furthermore, set $\tilde{A}_3 := \{(\tilde{P}_1, Q_2) \text{ not left coprime}\}$. From previous computations (see Theorem 12), one knows $\mathrm{Pr}(A_3) = O(t \cdot \mathrm{Pr}(\tilde{A}_3))$ as well as $\mathrm{Pr}(\tilde{A}_3) = t^{-1} \cdot \mathrm{Pr}(A_3) + O(\mathrm{Pr}(A_3))$. It follows

$$\mathrm{Pr}_{M_{p+m}=0}(A_{03}) = O\left(\mathrm{Pr}(A_3 \cap \{M_{p+m} = 0\})\right) = O(t \cdot \mathrm{Pr}(\tilde{A}_3 \cap \{M_{p+m} = 0\})) =$$
$$= O(t^2 \cdot \mathrm{Pr}(\tilde{A}_3)) = O(t \cdot \mathrm{Pr}(A_3)).$$

Here, for the second equation, one uses that the elements of p_{c_1} are not contained in M_{p+m} (and hence, a condition on M_{p+m} could not influence p_1) and that the additional condition one has considering A_3 instead of $\tilde{A}_3$ only consists of a restriction for p_1. The third equation follows per induction.

In summary, the probability of reachability is equal to

$$1 - \frac{\mathrm{Pr}(A_3) + O(t \cdot \mathrm{Pr}(A_3))}{1 + O(t)} = 1 - \frac{t^m + O(t^{m+1})}{1 + O(t)} = 1 - t^m + O(t^{m+1}).$$

$\square$

Corollary 16.
The probability that a feedback interconnection of two minimal systems is observable is equal to $1 - t^m + O(t^{m+1})$.

Proof.
Similar to the computation of observability for a series connection of two systems, the statement follows from the computation of the probability for reachability. Even with the condition $\det(I - D_2 D_1) \neq 0$ could be dealt in a similar way. Finally, note that the number of outputs of the second system p_2 equals the number of inputs for the first system m. $\square$

Remark 21.
If one considers a circular interconnection of N minimal systems with strictly proper transfer functions, the probability of reachability is equal to the probability that a series connection of N minimal systems with strictly proper transfer functions is reachable. This is due to the fact that in the strictly proper case, $I - D_N \cdots D_1 = I$ since $D_1 = \cdots = D_N = 0$ and therefore, $A_0 = \emptyset$.

3.5 Partial State Reachability of a Series Connection of three Systems

In [40], criteria for so-called partial state reachability of a series connection of reachable strictly proper systems were provided. The notion of partial state reacha- bility means that one is only interested in controling a subset of the node systems. In this section, we focus on a special case also considered in [40], namely a se- ries connection of three strictly proper single-input single-output (SISO) systems $(A_i, B_i, C_i) \in \mathbb{F}^{n_i \times n_i} \times \mathbb{F}^{n_i \times m_i} \times \mathbb{F}^{p_i \times n_i}$ for $i = 1, 2, 3$ and so called $(1, 3)$-reachability, i.e. it should be possible to steer the first and third system to arbitrary states while the state of the second system is of no importance. To be able to use the theorems provided in [40], we firstly, have to introduce and adopt some notation from there.

Definition 14.
*A series connection of three systems is called $(1, 3)$-**reachable** if for each $\xi_1 \in \mathbb{F}^{n_1}$ and each $\xi_3 \in \mathbb{F}^{n_3}$, there exist $\tau_* \in \mathbb{N}_0$ and a sequence of inputs $u(0), \dots, u(\tau_*) \in \mathbb{F}^m$ such that $x_i(\tau_* + 1) = \xi_i$ for $i = 1, 3$.*

Definition 15.
*For $n_f \in \mathbb{N}$ and $f_j \in \mathbb{F}$ for $j = -\infty, \dots, n_f$, $f(z) = \sum_{j=-\infty}^{n_f} f_j z^j$ is called **trun- cated Laurent series** and the vector space of all truncated Laurent series is denoted by $\mathbb{F}((z^{-1}))$. Moreover, we call $\pi : \mathbb{F}((z^{-1})) \to \mathbb{F}[z], f(z) \mapsto \sum_{j=0}^{n_f} f_j z^j$ the **projection onto the polynomial part**.*

With these notations, we are now able to state the corresponding theorem from [40] for the special case of single-input single-output.

Theorem 43. *[40, Theorem 8]*
The series connection of three reachable strictly proper SISO systems with transfer functions $C_i(zI - A_i)^{-1}B_i = \frac{p_i(z)}{q_i(z)} \in \mathbb{F}(z)$ and $\gcd(p_i, q_i) = 1$ for $i = 1, 2, 3$ is $(1, 3)$- reachable if and only if $\mathbb{F}[z] = \delta(z)\mathbb{F}[z] + \pi(d(z)\mathbb{F}[z])$, where $\delta := \gcd(q_3, p_1 p_2)$ and $d := \frac{p_1 p_2}{q_2}$.

To make the criterion from the preceding theorem easier to handle, we want to transform this criterion into a criterion on the coefficients of d and δ.

Theorem 44.
Write $\delta(z) = \sum_{j=0}^{n_\delta} \delta_j z^j$ and $d(z) = \sum_{j=-\infty}^{n_d} d_j z^j$ with $n_\delta \in \mathbb{N}_0$, $n_d \in \mathbb{Z}$ and $\delta_j, d_j \in \mathbb{F}$, where $\delta_{n_\delta} = 1$ and $d_{n_d} \neq 0$. Moreover, set $\delta_j = 0$ for $j > n_\delta$ and $d_j = 0$ for $j > n_d$. The series connection of three reachable strictly proper SISO systems is $(1, 3)$-reachable if and only if either $n_\delta = 0$ or $n_d \leq 0$ or $n_\delta, n_d \geq 1$ and for each $n \in \mathbb{N}_0$, there exist

$n_1, n_2 \in \mathbb{N}_0$ *such that the matrix*

$$
M := \begin{bmatrix}
d_0 & \cdots & d_{-n_2} & \delta_0 & & 0 \\
\vdots & \ddots & \vdots & \vdots & \ddots & \\
d_{n-n_2} & & d_0 & \delta_{n-n_1} & & \delta_0 \\
\vdots & \ddots & \vdots & \vdots & \ddots & \vdots \\
d_n & \cdots & d_{n-n_2} & \delta_n & \cdots & \delta_{n-n_1}
\end{bmatrix} \in \mathbb{F}^{(n+1) \times (n_1+n_2+2)}
$$

has full row rank.

Proof.
According to the preceding theorem, the considered interconnection is $(1,3)$-reachable if and only if for each $h \in \mathbb{F}[z]$ there exist $f, g \in \mathbb{F}[z]$ with $h = \delta f + \pi(dg)$. If $n_\delta = 0$, i.e. $\delta \equiv 1$, this clearly is possible; simply choose $f \equiv h$ and $g \equiv 0$. If $n_d \leq 0$, i.e. d is a proper rational function, it holds $\pi(d\,\mathbb{F}[z]) = \mathbb{F}[z]$ (see beginning of the proof of Corollary 9 in [40]) and the claim follows by choosing $f \equiv 0$. Hence, it only remains to consider the case $n_\delta, n_d \geq 1$. Therefore, write $h(z) = \sum_{j=0}^{n} h_j z^j$, $f(z) = \sum_{j=0}^{n_1} f_j z^j$ and $g(z) = \sum_{j=0}^{n_2} g_j z^j$ with $n, n_1, n_2 \in \mathbb{N}_0$ and $h_j, f_j, g_j \in \mathbb{F}$, where $h_n, f_{n_1}, g_{n_2} \neq 0$. Then, $(1,3)$-reachability is equivalent to the fact that for each $n \in \mathbb{N}_0$ and each $h \in \mathbb{F}[z]$ with $\deg(h) = n$, there exist $n_1, n_2 \in \mathbb{N}_0$ and $f, g \in \mathbb{F}[z]$ with $\deg(f) = n_1$ and $\deg(g) = n_2$ such that

$$
h_i = \sum_{j=0}^{i} \delta_j f_{i-j} + \sum_{j=-\infty}^{i} d_j g_{i-j} \quad \text{for} \quad 0 = 1, \ldots, n,
$$

which is equivalent to

$$
\begin{pmatrix} h_0 \\ \vdots \\ h_n \end{pmatrix} = \begin{bmatrix}
d_0 & \cdots & d_{-n_2} & \delta_0 & & 0 \\
\vdots & \ddots & \vdots & \vdots & \ddots & \\
d_{n-n_2} & & d_0 & \delta_{n-n_1} & & \delta_0 \\
\vdots & \ddots & \vdots & \vdots & \ddots & \vdots \\
d_n & \cdots & d_{n-n_2} & \delta_n & \cdots & \delta_{n-n_1}
\end{bmatrix} \begin{pmatrix} g_0 \\ \vdots \\ g_{n_2} \\ f_0 \\ \vdots \\ f_{n_1} \end{pmatrix}.
$$

Thus, $(1,3)$-reachability is equivalent to the fact that for each $n \in \mathbb{N}_0$, there exist $n_1, n_2 \in \mathbb{N}_0$ such that the linear map defined by M is surjective and the proof is complete. $\qquad\square$

With the help of the preceding theorem, we could investigate what happens in the special case that the second and third node system are scalar and the first node system is of dimension less than three.

Theorem 45.
If $n_2 = n_3 = 1$ and $n_1 \leq 2$, the series connection of three reachable strictly proper SISO systems is always $(1,3)$-reachable.

Proof.
It holds $n_\delta = \deg(\delta) \leq \deg(q_3) = n_3 = 1$. If $n_\delta = 0$, the statement follows directly from the preceding theorem. Therefore, one only has to consider the case $\deg(\delta) = 1 = \deg(q_3)$ and hence, $\delta \equiv q_3$. Since $\frac{p_1 p_2}{q_2} - \pi(d)$ is strictly proper, one has $p_1 p_2 = q_2 \cdot \pi(d) + r(d)$ with $\deg(r(d)) < \deg(q_2) = n_2 = 1$, i.e. $r(d)$ is constant.
Assume that the interconnection is not $(1,3)$-reachable, i.e. $n_d \geq 1$ and there exists $n \in \mathbb{N}_0$ such that for each $n_1, n_2 \in \mathbb{N}_0$, the matrix M (defined as in the preceding theorem) has no full row rank. Especially, for $n_1 = n_2 = 0$, if $n \geq 1$, $M = \begin{bmatrix} d_0 & \delta_0 \\ & 1 \\ \vdots & 0 \\ & \vdots \\ d_n & 0 \end{bmatrix}$

has no full row rank and if $n = 0$, $M = [d_0 \ \delta_0]$ has no full row rank. For $n \geq 1$, this implies $d_2 = \ldots = d_n = 0$, i.e. $n_d = 1$, as well as $d_0 - d_1 \delta_0 = 0$, i.e. $-\delta_0$ is a zero of $\pi(d)$. For $n = 0$, it follows $d_0 = \delta_0 = 0$ and hence, $-\delta_0 = 0$ is a zero of $\pi(d)$, too. Since $\delta(z) = z + \delta_0$ is the minimal polynomial of $-\delta_0$, one has in either case that $\delta \mid \pi(d)$. Because $\delta \mid p_1 p_2$ per definition, it follows $\delta \mid r(d)$, which implies $r(d) = 0$ since we already saw that $r(d)$ is constant. But then, $\pi(d) = d = \frac{p_1 p_2}{q_2}$ is a polynomial of degree at least one since it is divisible by δ. Since p_2/q_2 is strictly proper, this implies that p_1 is of degree at least two, a contradiction to $\deg(p_1) < n_1 \leq 2$. Consequently, such an interconnection has always to be $(1,3)$-reachable. $\qquad\square$

3.6   General Networks

In this section, we want to investigate what general statements are possible if one does not know the interconnection matrices K, L (and M), defined as at the beginning of Section 1.2. As first result one has that the probability of reachability either is zero or tends to one if the size of the field tends to infinity.

Theorem 46.
Let K, L be arbitrary but fixed matrices. The probability that $(\mathcal{A}, \mathcal{B})$ is reachable if (A_i, B_i, C_i, D_i) are chosen randomly for $i = 1, \ldots, N$, is either equal to zero or $1 + O(t)$ for $t \to 0$.

Proof.
One has to show that if there is a choice of (A_i, B_i, C_i, D_i) for $i = 1, \ldots, N$ such that $(\mathcal{A}, \mathcal{B})$ is reachable, than $(\mathcal{A}, \mathcal{B})$ is reachable with probability $1 + O(t)$ for random choice of (A_i, B_i, C_i, D_i). Let (A_i, B_i, C_i, D_i) for $i = 1, \ldots, N$ such that $(\mathcal{A}, \mathcal{B})$ is reachable. Then, there is a subminor of $\mathcal{R}(\mathcal{A}, \mathcal{B})$ that is nonzero. Consequently, if you view this subminor as rational function in the entries of (A_i, B_i, C_i, D_i) with coefficients depending on the entries of K and L, the numerator of this rational function is not the zero polynomial. Hence, according to Lemma 3, the probability that this subminor is equal to zero is $O(t)$. Hence, the probability that $\mathcal{R}(\mathcal{A}, \mathcal{B})$ is not of full row rank is $O(t)$ as well, and we are done. $\qquad\square$

In order to describe for which matrices K and L the probability of reachability is equal to zero and in which cases it tends to one, we need the following definition.

Definition 16.
In the following, we view the N node systems of a network of linear systems as vertices of a graph. Moreover, we add a $N + 1$-th vertex, called input node. We denote by Γ_{KL} the directed graph where for $i, j \in \{1, \ldots, N\}$, there is an edge from vertex j to vertex i if and only if $K_{ij} \in \mathbb{F}^{p_j \times m_i}$ is not the zero matrix and for $i = 1, \ldots, N$, there is an edge from the input node to vertex i if and only if $L_i \in \mathbb{F}^{m_i \times m}$ is not the zero matrix. In particular, the input node has no ingoing edges. Furthermore, we assume that Γ_{KL} contains no self-loops, i.e. $K_{ii} \equiv 0$ for $i = 1, \ldots, N$. A vertex of Γ_{KL} is called accessible if the graph contains a path from the input node to this vertex. Otherwise, it is called nonaccessible.

For the proof of the following theorem, we will additionally need the following definition.

Definition 17.
*For an arbitrary field $\mathbb{F}$, a matrix $A \in \mathbb{F}^{n \times n}$ is called **structured** if its entries a_{ij} are either fixed zeros or free variables from $\mathbb{F}$.*
One defines the corresponding graph $\Gamma_A = (V, E_A)$, such that $V = \{1, \ldots, N\}$ and $E_A = \{(i, j) \in V \times V \mid a_{ji} \text{ is a free variable}\}$.
One can extend this definition to matrix pairs $(A, B) \in \mathbb{F}^{n \times n} \times \mathbb{F}^{n \times m}$ by setting $\Gamma_{(A,B)} = \Gamma_{[A\ B]}$. In this case, the last m vertices can only have outgoing edges, which is consistent with regarding them as inputs of the system (A, B).
One calls the pair (A, B) structural controllable if there exists a choice for the free parameters such that one has reachability.

Theorem 47.

    (a) *If Γ_{KL} contains a nonaccessible vertex, $(\mathcal{A}, \mathcal{B})$ is not reachable, i.e. the probability of reachability is equal to zero.*

    (b) *If all vertices of Γ_{KL} are accessible, there exists $s \in \mathbb{N}$ such that $(\mathcal{A}, \mathcal{B})$ is reachable over the extension field $\mathbb{F}^r$ of $\mathbb{F}$ with probability $1 + O(t^r)$ if $r \geq s$ and with probability zero if $r < s$.*

Proof.

    (a) Follows obviously from the definition of K, L and Γ_{KL}.

    (b) For this part of the proof, we need some theory of structural controllability over the field $\mathbb{R}$, which could be found in [28], [29], [37] or [16]. It follows from these papers that there is no choice for the free parameters in A and B such that the pair $(A, B) \in \mathbb{R}^{n \times n} \times \mathbb{R}^{n \times m}$ is reachable if and only if $\Gamma_{(A,B)}$ has a nonaccessible node or $[A\ B]$ has no full row rank for every choice of the free parameters in A and B. Moreover, if one considers the proofs for this statement, one could observe that they are also valid for a finite field $\mathbb{F}$ if its size is large enough (essentially it has to be possible to choose the free parameters in such way that the eigenvalues of certain matrices are distinct). Hence, we could apply this statement to the pair (A, B), which is structured by fixing K and L and viewing the entries of (A, B, C, D) as free variables. Furthermore the case that $[A\ B]$ has no full row rank for every choice of (A, B, C, D) could be excluded since the choice of $A = I$ and $B = D = 0$ leads to $[A\ B] = [I\ 0]$, which clearly has full row rank independently of K and L. It remains to show that if $\Gamma_{[A\ B]}$ contains a nonaccessible node, then Γ_{KL} contains a nonaccessible node. In the papers concerning structural controllability over $\mathbb{R}$ mentioned above, it is stated that $\Gamma_{[A\ B]}$ contains a nonaccessible node if and only if it

is of the form $A = \begin{bmatrix} A_{11} & 0_{k\times(\bar{n}-k)} \\ A_{21} & A_{22} \end{bmatrix}$, $B = \begin{pmatrix} 0_{k\times m} \\ B_{22} \end{pmatrix}$ for some $1 \le k < \bar{n}$.

Since the entries of (A, B, C, D) are free variables, one could assume $D = 0$.

In this case, one has $[A\ B] = \begin{bmatrix} A_1 & & B_1K_{1N}C_N & B_1L_1 \\ & \ddots & & \vdots \\ B_NK_{N1}C_1 & & A_N & B_NL_N \end{bmatrix}$. That

K and L are such that this matrix is of the nonaccessible form for all choices of A and B, implies $L_1 = \cdots = L_r = 0$ for some integer $1 \le r \le N$ as well as $K_{1j} = \cdots = K_{rj} = 0$ for $j \ge r + 1$. Hence, for $i \le r$, there is no path in Γ_{KL} from the input node to vertex i and thus, Γ_{KL} contains a nonaccessible vertex. □

Remark 22.
Structural controllability over a finite field $\mathbb{F}$ has already been studied in [38]. But there, it was assumed that B is fixed (consisting of pairwise different unit vectors) and only A is structured. Under these assumption, for several structures of A, lower bounds for the required field size to achieve structural controllabiliy were given.

3.7 Homogeneous Networks

In this section, we somehow reverse the question under investigation. So far, we fixed the interconnection matrices K, L and M and chose the node systems randomly. Now, we want to choose the interconnection matrices randomly. This is easily possible for so-called homogeneuos networks since - as we will see - the choice of the node systems does not influence the reachability of such a network.

Definition 18.
*A network is called **homogeneous** if the node systems are identical minimal strictly proper single-input single-output (SISO) systems, i.e. $(A_i, b_i, c_i) = (A, b, c) \in \mathbb{F}^{n\times n} \times \mathbb{F}^n \times \mathbb{F}^{1\times n}$ for $i = 1, \ldots, N$ with (A, b, c) reachable and observable.*

Theorem 48. *[14], [19]*
For homogeneous networks, $(\mathcal{A}, \mathcal{B}, \mathcal{C})$ is reachable/observable if and only if (K, L, M) is reachable/observable. In particular, reachability/observability only depends on the interconnection structure and not on (A, b, c).

Since all node systems of a homogeneuos network are SISO, one has $p_i = m_i = 1$ and consequently, $K_{ij} \in \mathbb{F}$ and $L_i \in \mathbb{F}^{1 \times m}$ for $i, j \in \{1, \ldots, N\}$.

Combining the previous theorem with the probability estimations of the preceding sections, one could compute the probabilities of reachability, observability and minimality for a homogenous network with randomly chosen interconnection structure.

Corollary 17.
For random $(K, L, M) \in \mathbb{F}^{N \times N} \times \mathbb{F}^{N \times m} \times \mathbb{F}^{p \times N}$, the probability that a homogeneous network (consisting of N SISO systems) is

(i) reachable is equal to $\prod_{j=m}^{N+m-1}(1 - t^j) = 1 - t^m + O(t^{m+1})$,

(ii) observable is equal to $\prod_{j=p}^{N+p-1}(1 - t^j) = 1 - t^p + O(t^{p+1})$,

(iii) minimal is equal to $1 - t^p - t^m + O(t^{\min(m,p)+1})$.

Next, structured matrix pairs (K, L) should be considered. For the case that one could partition the node systems into blocks which cannot influence each other (i.e. a system could only influence other systems that are in the same block), one gets structured matrices K and L as in the following corollary:

Corollary 18.
Let $(K_i, L_i) \in \mathbb{F}^{n_i \times n_i} \times \mathbb{F}^{n_i \times m_i}$ for $i = 1, \ldots, r$ with $N = n_1 + \cdots + n_r$ and $m = m_1 + \cdots + m_r$ be randomly. Then, the probability that the homogeneus network with

$$
K = \begin{bmatrix} K_1 & & 0 \\ & \ddots & \\ 0 & & K_r \end{bmatrix}, \quad L = \begin{bmatrix} L_1 \\ \vdots \\ L_r \end{bmatrix}
$$

is reachable is equal to

$$
1 - \sum_{k=1}^{m+1} \binom{r}{k} t^m + O\left(t^{m+1}\right).
$$

Proof.
The theorem follows from the fact that the sought-after probability is equal to the probability that the parallel connection of random matrix pairs $(K_i, L_i) \in \mathbb{F}^{n_i \times n_i} \times \mathbb{F}^{n_i \times m_i}$ for $i = 1, \ldots, r$ is reachable. $\qquad \square$

3.8 Conclusion

In this chapter, we used probability estimations for coprimeness conditions on polynomial matrices to calculate the probabilities of reachability, observability and minimality for different kinds of networks of linear systems. For many of the computations, we could resort to results from the preceding chapter or could at least adapt counting methods from there.

However, linear systems theory is not the only field, where polynomial matrices are of interest. For example, especially polynomial matrices over finite fields play an important role in coding theory, in particular for the investigation of convolutional codes. However, since convolutional codes are closely related to linear systems, in most of the applications contained in the following chapter, we do not need to go back to polynomial matrices but could transfer the results on networks of linear systems quite directly to interconnected convolutional codes. This fact, was important for the choice of the interconnection structures which we considered throughout this chapter. The networks, we focused on are not only some standard examples of networks but could also be found in the coding literature used to concatenated convolutional codes, see e.g. [2], [3], [7], [9], [12], [24].

Chapter 4

Applications to Coding Theory

The results concerning polynomial matrices from Chapter 2 as well as those concerning reachability and observability of linear systems from Chapter 3 could be transferred to probability calculations in the area of coding theory.

Referring to linear systems the most obvious connection could be drawn to convolutional codes since one could define these codes in terms of a linear system. When considering block codes, no polynomial matrices occur but only constant matrices. Although this implicates that it is not possible to transfer results from the preceding chapters directly, one could use similar counting methods to calculate probabilities of important properties in this field.

In the first section of this chapter, we introduce the basic notions and statements relating to the theory of block and convolutional codes.

In Section 4.2, it is explained how one could construct a convolutional code out of a linear system and what are the implications of this linear system representation of the corresponding code.

In Section 4.3, we consider one convolutional code and calculate the probability that it is non-catastrophic, which is a crucial property of such a code.

In Section 4.4, we look at interconnections of several convolutional codes. We do this with the help of Theorem 10, which we transfer to the situation of convolutional codes. In the various subsections, we consider different examples of interconnected convolutional codes occurring in the coding literature. We calculate their probabilities of being represented by a linear system in a minimal way and moreover, the probability of being non-catastrophic.

In Section 4.5, we investigate another type of interconnected convolutional code but now, with one of the two constituent codes being a block code and the other being a convolutional code. After considering the non-catastrophicity of such a code, we estimate the probability that it is of maximum distance profile.

In the following Section 4.6, we continue this examination by computing a lower and an upper bound for the probability that a general convolutional code is of maximum distance profile.

In Section 4.7, we do the analogous consideration for block codes, i.e. derive bounds for the probability that a block code is MDS.

Finally, in Section 4.8. we take a look at random linear network coding and calculate the probability of finding a solution for a network coding problem, at first for the case of delay-free acyclic network coding and then for the case of memory-free convolutional network coding.

4.1 Definitions and Basics about Block and Convolutional Codes

This section gives a short overview about the definitions and properties concerning block and convolutional codes to lay the foundations for the calculations of the following sections. Because—as we will see later—block codes could be viewed as a special type of convolutional codes, we will start with defining and investigating block codes.

Definition 19.
*A $[\mathbf{n}, \mathbf{k}]$-**block code** $\mathcal{B}$ is a k-dimensional subspace of $\mathbb{F}^n$, i.e. there exists $G \in \mathbb{F}^{n \times k}$ of full column rank such that*

$$\mathcal{B} = \{v \in \mathbb{F}^n \mid v = Gu \text{ for some } u \in \mathbb{F}^k\}.$$

*G is called **generator matrix** of the code and is unique up to right multiplication with an invertible matrix $U \in Gl_k(\mathbb{F})$. Furthermore, $u \in \mathbb{F}^k$ is called **message vector** and the elements $v \in B$ are called **codewords**.*

That G is of full column rank implies $n \geq k$, i.e. a codeword contains redundancy and thus, more information than the original message to enable error detection and correction. Important for the performance of a code in terms of error-free decoding is the minimum distance between two codewords. Since we consider linear codes, this value could be expressed by the Hamming weight of the codewords.

Definition 20.
*The **Hamming weight wt(v)** of $v \in \mathbb{F}^n$ is defined as the number of its nonzero components.*

Definition 21.
*The **minimum distance** of a block code $\mathcal{B}$ is defined as*

$$d_{min}(\mathcal{B}) = \min_{v \in \mathcal{B}}\{wt(v) \mid v \neq 0\}.$$

It is desirable to achieve a minimum distance that is as large as possible. However, this means that one also needs a large number of codewords and/or sufficiently long codewords. This fact is implied in the following theorem.

Theorem 49. *(Singleton bound)*
For every $[n, k]$-block code $\mathcal{B}$, it holds $d_{min}(\mathcal{B}) \leq n - k + 1$.

Surely, a code will in some sense have optimal parameters if it attains the Singleton bound, which draws particular interest to this kind of codes.

Definition 22.
*A $[n, k]$-block code $\mathcal{B}$ is called **maximum distance separable (MDS)** if $d_{min}(\mathcal{B}) = n - k + 1$.*

For our probability estimations in later parts of this chapter, we will use the following characterization of MDS codes in terms of the corresponding generator matrix.

Theorem 50. *[8, Theorem 1]*
Let $G = \begin{pmatrix} I_k \\ P \end{pmatrix} \in \mathbb{F}^{n \times k}$ be a generator matrix for the block code $\mathcal{B}$. Then, $\mathcal{B}$ is a MDS block code if and only if every square submatrix of P is nonsingular.

Next, we want to consider so-called convolutional codes, which are obtained if one works over the ring of polynomials $\mathbb{F}[z]$ instead of working over $\mathbb{F}$, as done for block codes.

Definition 23.
*A **convolutional code** $\mathcal{C}$ of **rate** k/n is a free $\mathbb{F}[z]$-submodule of $\mathbb{F}[z]^n$ of rank k. Hence, there exists $G \in \mathbb{F}[z]^{n \times k}$ of full column rank such that*

$$\mathcal{C} = \{v \in \mathbb{F}[z]^n \mid v(z) = G(z)m(z) \text{ for some } m \in \mathbb{F}[z]^k\}.$$

*G is called **generator matrix** of the code and is unique up to right multiplication with a unimodular matrix $U \in Gl_k(\mathbb{F}[z])$.*

In contrast to block codes, one has a third parameter characterizing a convolutional code, namely the so-called degree of the code.

Definition 24.
*Let $\nu_1, \ldots, \nu_k$ be the column degrees of $G \in \mathbb{F}[z]^{n \times k}$. Then, $\nu := \nu_1 + \cdots + \nu_k$ is called the **order** of G. The **degree** δ of a convolutional code $\mathcal{C}$ is defined as the minimal order of its generator matrices. Equivalently, one could define the degree of $\mathcal{C}$ as the maximal degree of the $k \times k$-minors of one and hence, each generator matrix of $\mathcal{C}$.*

Theorem 51.
It holds $\nu = \delta$, i.e. G is a minimal basis of $\mathcal{C}$, if and only if G is column proper.

Next, we define a very important property of a convolutional code, namely the so-called non-catastrophicity. This quality of a code will be crucial for our later considerations since its probability could be deduced from former calculations of this dissertation.

Definition 25.
*A convolutional code $\mathcal{C}$ is called **non-catastrophic** if one and therefore, each of its generator matrices is right prime.*

As for block codes it is important to consider distance properties of convolutional codes.

Definition 26.
*For $v \in \mathbb{F}[z]^n$ with $\deg(v) = \gamma$, write $v(z) = v_0 z^\gamma + \cdots + v_\gamma$ with $v_\tau \in \mathbb{F}^n$ for $\tau = 0, \ldots, \gamma$ and set $v_\tau = 0 \in \mathbb{F}^n$ for $\tau \geq \gamma + 1$. Then, for $j \in \mathbb{N}_0$, the **j-th column distance** of a convolutional code $\mathfrak{C}$ is defined as*

$$d_j^C(\mathfrak{C}) := \min_{v \in \mathfrak{C}} \left\{ \sum_{t=0}^{j} wt(v_t) \mid v \neq 0 \right\}.$$

*Moreover, $d_{free}(\mathfrak{C}) := \lim_{j \to \infty} d_j^C(\mathfrak{C})$ is called the **free distance** of $\mathfrak{C}$.*

There exists an analogue to the Singleton bound for convolutional codes as stated in the following theorem.

Theorem 52. *[34] (Generalized Singleton bound)*
For a convolutional code $\mathfrak{C}$ of rate k/n and degree δ, it holds

$$d_{free}(\mathfrak{C}) \leq (n-k) \left(\left\lfloor \frac{\delta}{k} \right\rfloor + 1 \right) + \delta + 1.$$

Again, we are interested in codes with good distance properties but now not only in those reaching the generalized Singleton bound but also in those whose column distances are optimal.

Definition 27. *[25]*
A convolutional code $\mathfrak{C}$ of rate k/n and degree δ is called

(i) **maximum distance separable (MDS)** *if*

$$d_{free}(\mathfrak{C}) = (n-k) \left(\left\lfloor \frac{\delta}{k} \right\rfloor + 1 \right) + \delta + 1,$$

(ii) *of* **maximum distance profile (MDP)** *if*

$$d_j^C(\mathfrak{C}) = (n-k)(j+1) + 1 \quad \text{for } j = 0, \ldots, L \quad \text{with } L = \left\lfloor \frac{\delta}{k} \right\rfloor + \left\lfloor \frac{\delta}{n-k} \right\rfloor,$$

(iii) **strongly MDS** *if*

$$d_M^C(\mathfrak{C}) = (n-k) \left(\left\lfloor \frac{\delta}{k} \right\rfloor + 1 \right) + \delta + 1 \quad \text{with} \quad M = \left\lfloor \frac{\delta}{k} \right\rfloor + \left\lceil \frac{\delta}{n-k} \right\rceil.$$

We conclude this section with possibilities to check the distances properties of a convolutional code.

Theorem 53. *[11, Theorem 1.3]*
Let $G(z) = \sum_{i=1}^{\mu} G_i z^i \in \mathbb{F}[z]^{n \times k}$ be the generator matrix of a convolutional code with degree δ and define $\mathcal{G} = \begin{bmatrix} G_0 & & 0 \\ \vdots & \ddots & \\ G_L & \cdots & G_0 \end{bmatrix}$ where $G_i \equiv 0$ for $i > \mu$. Then, the corresponding convolutional code is of maximum distance profile if and only if every full size minor of $\mathcal{G}$ that is not trivially zero, i.e. zero for all choices of $G_1, \ldots, G_L$, is nonzero.

Remark 23. *[8]*

(i) *Every strongly MDS code is a MDS code.*

(ii) *If $n - k$ divides δ, a convolutional code $\mathfrak{C}$ has maximum distance profile if and only if $\mathfrak{C}$ is strongly MDS.*

4.2 Correspondence between Linear Systems and Convolutional Codes

The aim of this section is to explain how one could construct a convolutional code based on a linear system (see [35]). To this end, we start with a linear system $(A, B, C, D) \in \mathbb{F}^{s \times s} \times \mathbb{F}^{s \times k} \times \mathbb{F}^{(n-k) \times s} \times \mathbb{F}^{(n-k) \times k}$ and define

$$H(z) := \begin{bmatrix} zI - A & 0_{s \times (n-k)} & -B \\ -C & I_{n-k} & -D \end{bmatrix}.$$

The set of $\begin{pmatrix} y \\ u \end{pmatrix} \in \mathbb{F}[z]^n$ with $y \in \mathbb{F}[z]^{n-k}$ and $u \in \mathbb{F}[z]^k$ for which there exists $x \in \mathbb{F}[z]^s$ with $H(z) \cdot [x(z)\ y(z)\ u(z)]^\top = 0$ forms a submodule of $\mathbb{F}[z]^n$ of rank k and thus, a convolutional code of rate k/n, which is denoted by $\mathfrak{C}(A, B, C, D)$.

Moreover, if one writes $x(z) = x_0 z^\gamma + \cdots + x_\gamma$, $y(z) = y_0 z^\gamma + \cdots + y_\gamma$ and $u(z) = u_0 z^\gamma + \cdots + u_\gamma$ with $\gamma = \max(\deg(x), \deg(y), \deg(u))$, it holds

$$x_{\tau+1} = Ax_\tau + Bu_\tau$$
$$y_\tau = Cx_\tau + Du_\tau$$
$$(x_\tau, y_\tau, u_\tau) = 0 \text{ for } \tau > \gamma.$$

Furthermore, there exist $X \in \mathbb{F}[z]^{s \times k}, Y \in \mathbb{F}[z]^{(n-k) \times k}, U \in \mathbb{F}[z]^{k \times k}$ such that $\ker(H(z)) = \operatorname{im}[X(z)^\top\ Y(z)^\top\ U(z)^\top]^\top$ and $G(z) = \begin{pmatrix} Y(z) \\ U(z) \end{pmatrix}$ is a generator matrix for $\mathfrak{C}$ with $C(zI - A)^{-1}B + D = Y(z)U(z)^{-1}$.

Conversely, for each convolutional code $\mathfrak{C}$ of rate k/n and degree δ, there exists $(A, B, C, D) \in \mathbb{F}^{s \times s} \times \mathbb{F}^{s \times k} \times \mathbb{F}^{(n-k) \times s} \times \mathbb{F}^{(n-k) \times k}$ with $s \geq \delta$ such that $\mathfrak{C} = \mathfrak{C}(A, B, C, D)$. Moreover, it is always possible to choose $s = \delta$. In this case, one calls (A, B, C, D) a **minimal representation** of $\mathfrak{C}$.

The following two theorems from [35] are of special interest for us since they show how reachability and observability of linear systems, which we investigated in the preceding chapter, are related to properties of the corresponding convolutional code.

Theorem 54. *[35]*
(A, B, C, D) is a minimal representation of $\mathfrak{C}(A, B, C, D)$ if and only if it is reachable.

Theorem 55. *[35]*
Assume that (A, B, C, D) is reachable. Then $\mathfrak{C}(A, B, C, D)$ is non-catastrophic if and only if (A, B, C, D) is observable.

Since a convolutional code might have different realizations, partly minimal and partly not, we will need the following theorem to be able to compute the probability of non-catastrophicity for a convolutional code.

Theorem 56.
If (A, B, C, D) is a minimal representation of a convolutional code $\mathfrak{C}$, the set of all minimal representations of $\mathfrak{C}$ is given by $\{(SAS^{-1}, SB, CS^{-1}, D) \mid S \in Gl_\delta(\mathbb{F})\}$.

Proof.
Clearly, $(SAS^{-1}, SB, CS^{-1}, D)$ is a minimal representation of $\mathfrak{C}$. On the other hand, let (A, B, C, D) and $(\tilde{A}, \tilde{B}, \tilde{C}, \tilde{D})$ be minimal representations of $\mathfrak{C}$. Set $K := \begin{pmatrix} -I \\ 0 \end{pmatrix}$, $L := \begin{pmatrix} A \\ C \end{pmatrix}$ and $M := \begin{bmatrix} 0 & B \\ -I & D \end{bmatrix}$ and define $\tilde{K}$, $\tilde{L}$ and $\tilde{M}$ analogously. It follows from Theorem 3.4 of [33] (or from Theorem 6.7 of [32]) that there exist (unique) invertible matrices S and T such that $(\tilde{K}, \tilde{L}, \tilde{M}) = (TKS^{-1}, TLS^{-1}, TM)$. Write $T = \begin{bmatrix} T_1 & T_2 \\ T_3 & T_4 \end{bmatrix}$. Thus, the first of the preceding equations, implies $T_1 = S$ and $T_3 = 0$. Inserting this into the second equation, leads to $\tilde{A} = SAS^{-1} + T_2CS^{-1}$ and $\tilde{C} = T_4CS^{-1}$. Finally, the third equation yields $T_2 = 0$, $T_4 = I$ and using this $\tilde{B} = SB$ as well as $\tilde{D} = D$. $\qquad\square$

In Theorem 53, we characterized the MDP property of a convolutional code in terms of its generator matrix. But it is also possible to describe this property using a linear system representing the code.

Theorem 57. *[11, Corollary 1.1]*
The matrices (A, B, C, D) generate a MDP convolutional code if and only if the matrix

$$\mathcal{F}_L := \begin{bmatrix} D & 0 & \cdots & 0 \\ CB & \ddots & \ddots & \vdots \\ \vdots & \ddots & \ddots & 0 \\ CA^{L-1}B & \cdots & CB & D \end{bmatrix}$$

has the property that every minor which is not trivially zero is nonzero.

Remark 24.

(i) *It is also possible to make the considerations of this section for the case $\delta = 0$. In this case, a minimal representation consists only of the matrix D because the matrices A, B and C have zero rows or columns, respectively, and hence, these matrices do not exist. If one uses the convention that non-existing matrices have every property, one has that the corresponding linear system is always minimal. This corresponds to the fact that a representation with $s = 0$ is always a minimal representation as well as to the fact that a generator matrix of a convolutional code with degree $\delta = 0$ is constant and of full column rank and therefore, always right prime. Moreover, one has $H = [I \ - \ D]$, and for every $U \in Gl_k(\mathbb{F})$ and*

$Y := DU$, it holds that $\begin{pmatrix} Y \\ U \end{pmatrix}$ is a generator matrix for the code with $D = YU^{-1}$. The corresponding linear system only consists of the equation $y_t = Du_t$.

(ii) Actually, the set of generator matrices for convolutional codes of degree 0 and rate $\frac{k}{n}$ coincides with the set of generator matrices for $[n, k]$-block codes. The first one is a submodule of $\mathbb{F}[z]^n$ and the second a subspace of $\mathbb{F}^n$ but the two are isomorphic. This fact will be of interest in Section 4.5.

4.3 Probability of Non-Catastrophicity for a Convolutional Code

With the help of the theorems from the previous section, it is possible to transfer the probability results for linear systems to probability results for convolutional codes. As in the preceding chapter, we firstly consider just one convolutional code. Since we have to avoid system matrices with zero rows or columns, in this and the following section, all occuring convolutional codes are assumed to have positive degree. Convolutional codes of degree 0, which are basically the same as block codes (see Remark 24 (ii)), are treated in Section 4.5.

Theorem 58.
The probability that a convolutional code $\mathfrak{C}(A, B, C, D)$ of rate k/n and degree $\delta \geq 1$ is non-catastrophic is equal to

$$P^{rc}_{n-k,\delta,k} = \frac{\Pr((A, B, C, D) \text{ reachable and observable})}{\Pr((A, B, C, D) \text{ reachable})} = \qquad (4.1)$$

$$= 1 - t^{n-k} + O(t^{n-k+1}). \qquad (4.2)$$

Proof.
Equations (4.1) and (4.2) are simply the statements from Lemma 13 and Theorem 12. Hence, it remains to show that the probability of non-catastrophicity is equal to one of the expressions from (4.1). Consequently, there are two possibilities to prove this theorem.
The first way is to show that the probability of non-catastrophicity is equal to $P^{rc}_{n-k,\delta,k}$. From the previous subsection, one knows that there exists
$(A, B, C, D) \in \mathbb{F}^{\delta \times \delta} \times \mathbb{F}^{\delta \times k} \times \mathbb{F}^{(n-k) \times \delta} \times \mathbb{F}^{(n-k) \times k}$ such that $\mathfrak{C} = \mathfrak{C}(A, B, C, D)$ and a generator matrix of $\mathfrak{C}$ of the form $G = \begin{pmatrix} Y \\ U \end{pmatrix}$ with $C(zI - A)^{-1}B + D = Y(z)U(z)^{-1}$.
Since G is of full column rank and unimodular equivalent generator matrices define the same convolutional code, one could assume that U is in Kronecker-Hermite form. In particular, it is column proper and because YU^{-1} is proper, it follows from Lemma 1 that $\deg_j(Y) \leq \deg_j(U)$ for $j = 1, \ldots k$. Finally, one knows from Theorem 51 that $\deg(\det(U)) = \delta$. Consequently, $G \in M(n - k, \delta, k)$ (see Definition 12) and since non-catastrophicity of $\mathfrak{C}$ is equivalent to right primeness of G, the statement follows. A second way to prove this theorem is to use Theorem 55 and Theorem 56. These

theorems imply that the probability of non-catastrophicity is equal to the right hand side of equation (4.1). □

Remark 25.

   (i) *Combining ways one and two of the proof for the preceding theorem, leads to an alternative proof for Lemma 13.*

  (ii) *The probability that a convolutional code of rate $1/n$ and degree $\delta \geq 1$ is non-catastrophic is (exactly) $1 - t^{n-1}$.*

 (iii) *With the same argumention as in Example 2 (i), one gets that the probability that a convolutional code of degree 1 and rate $\frac{k}{n}$ is non-catastrophic is (exactly) $1 - t^{n-k}$.*

 (iv) *Since each matrix from $\mathbb{F}^{n \times k}$ of full column rank is right prime, every convolutional code of degree 0 is non-catastrophic (see Remark 24 (i)).*

If one only fixes the rate of a convolutional code and chooses the degree randomly, i.e. there are no degree bounds for the polynomial entries of the generator matrix, the corresponding probability of non-catastrophicity could be expressed by the natural density.

Remark 26.

The natural density of a convolutional code of rate $\frac{k}{n}$ to be non-catastrophic is equal to $\prod_{j=n-k}^{n-1}(1 - t^j)$.

Proof.

The statement follows from Proposition 6 of [17], which states that the natural density that a square matrix has a constant nonzero determinant is equal to zero. Therefore, the natural density that all square submatrices of an arbitrary polynomial rectangular matrix have a determinant equal to zero is zero, too. Thus, the natural density that a generator matrix of a code, i.e. a matrix containing a square submatrix with determinant unequal to the zero polynomial, is right prime is equal to the natural density that an arbitrary polynomial matrix is right prime.

□

4.4 Interconnected Convolutional Codes

In this section, we want to consider interconnected convolutional codes. To this end, we start with transferring Theorem 10 to the situation of convolutional codes. Doing this, one achieves the following theorem:

Theorem 59.

Let $\mathcal{C}_i = \mathcal{C}(A_i, B_i, C_i, D_i)$ be non-catastrophic with (A_i, B_i, C_i, D_i) minimal for $i = 1, \ldots, N$. With the same notation as in Theorem 10, set

$$\mathcal{A} = A + BK(I - DK)^{-1}C$$
$$\mathcal{B} = B(I - KD)^{-1}L$$
$$\mathcal{C} = M(I - DK)^{-1}C$$
$$\mathcal{D} = M(I - DK)^{-1}DL + J$$

and consider coprime factorizations $C_i(zI - A_i)^{-1}B_i + D_i = P_i(z)Q_i(z)^{-1}$.
If $\det(I - DK) \neq 0$, *one has:*

1. (A, B, C, D) *is a minimal representation of* $\mathfrak{C}(A, B, C, D)$ *if and only if* $(Q(z) - KP(z), L)$ *are left coprime*

2. (A, B, C, D) *is a minimal representation of* $\mathfrak{C} = \mathfrak{C}(A, B, C, D)$ *and* $\mathfrak{C}$ *is non-catastrophic if and only if* $(Q(z) - KP(z), L)$ *are left coprime and* $(Q(z) - KP(z), MP(z))$ *are right coprime.*

In the following, this theorem should be applied to some interconnection structures used to connect convolutional codes. The first two interconnection models draw our interest since they were used in [7], where sufficient criteria for (A, B, C, D) to be a minimal representation of $\mathfrak{C}(A, B, C, D)$ were provided.

4.4.1 Series Connection of two Convolutional Codes

In this model, the interconnection matrices are

$$K = \begin{bmatrix} 0 & 0 \\ I_{k_2} & 0 \end{bmatrix}, \quad L = \begin{bmatrix} I_{k_1} \\ 0 \end{bmatrix}, \quad M = I_{n_2},$$

where $k_2 = n_1 - k_1$ because the output of the first input serves as input for the second system. Since the matrices K and L are the same as in Example 1 (b), and M is not relevant for the reachability of the interconnection, the corresponding probability is the same as in Theorem 34. Hence, one gets:

Theorem 60.
Let $(A_i, B_i, C_i, D_i) \in \mathbb{F}^{\delta_i \times \delta_i} \times \mathbb{F}^{\delta_i \times k_i} \times \mathbb{F}^{(n_i - k_i) \times \delta_i} \times \mathbb{F}^{(n_i - k_i) \times k_i}$ *be minimal representations of the non-catastrophic convolutional codes* $\mathfrak{C}(A_i, B_i, C_i, D_i)$ *for* $i = 1, 2$. *Then, the probability that* (A, B, C, D) *is a minimal representation for the series connected code* $\mathfrak{C}(A, B, C, D)$ *is equal to*

$$1 - t^{k_1} + O\left(t^{k_1+1}\right).$$

Since $M = I$, the observability of (A_i, B_i, C_i, D_i) obviously implies the observability of (A, B, C, D). This could also be seen using Theorem 10: $\begin{bmatrix} Q - KP \\ MP \end{bmatrix} =$

$\begin{bmatrix} Q_1 & 0 \\ -P_1 & Q_2 \\ P_1 & 0 \\ 0 & P_2 \end{bmatrix}$ is right prime if and only if $\begin{bmatrix} Q_1 & 0 \\ P_1 & 0 \\ 0 & Q_2 \\ 0 & P_2 \end{bmatrix}$ is right prime, which is true

since $\begin{bmatrix} Q_i \\ P_i \end{bmatrix}$ are right prime. Thus, one has:

Theorem 61.
Let $(A_i, B_i, C_i, D_i) \in \mathbb{F}^{\delta_i \times \delta_i} \times \mathbb{F}^{\delta_i \times k_i} \times \mathbb{F}^{(n_i - k_i) \times \delta_i} \times \mathbb{F}^{(n_i - k_i) \times k_i}$ *be minimal representations of the non-catastrophic convolutional codes* $\mathfrak{C}(A_i, B_i, C_i, D_i)$ *for* $i = 1, 2$. *Then,*

the probability that (A, B, C, D) *is a minimal representation for the series connected code* $\mathfrak{C}(A, B, C, D)$ *and that* $\mathfrak{C}(A, B, C, D)$ *is non-catastrophic is equal to*

$$1 - t^{k_1} + O\left(t^{k_1+1}\right).$$

Furthermore, the probability that this interconnection is minimal is equal to the probability that the single systems are minimal and P_1 and Q_2 are left coprime. According to Theorem 27 and Theorem 34, this probability is

$$\prod_{i=1}^{2}(1 - t^{k_i} - t^{n_i-k_i} + O(t^{\min(k_i,n_i-k_i)+1})) \cdot (1 - t^{k_1} + O(t^{k_1+1})) =$$

$$= 1 - 2t^{k_1} - 2t^{k_2} - t^{n_2-k_2} + O(t^{\min(k_1,k_2,n_2-k_2)+1})$$

since $n_1 - k_1 = k_2$. Hence, one has the following theorem:

Theorem 62.
Let $(A_i, B_i, C_i, D_i) \in \mathbb{F}^{\delta_i \times \delta_i} \times \mathbb{F}^{\delta_i \times k_i} \times \mathbb{F}^{(n_i-k_i) \times \delta_i} \times \mathbb{F}^{(n_i-k_i) \times k_i}$ *be randomly for* $i = 1, 2$. *Then, the probability that* (A, B, C, D) *is a minimal representation for the series connected code* $\mathfrak{C}(A, B, C, D)$ *and that* $\mathfrak{C}(A, B, C, D)$ *is non-catastrophic is equal to*

$$1 - 2t^{k_1} - 2t^{k_2} - t^{n_1-k_1} + O(t^{\min(k_1,k_2,n_2-k_2)+1}).$$

Remark 27.
In [7], where this interconnection structure was introduced, it is shown that $\mathrm{rk}(D_1) = k_1$ *is a sufficient condition to lower bound the free distance of the concatenated code by the free distance of the second code.*
According to Lemma 4, the probability that this condition is fulfilled is equal to $\prod_{j=n_1-2k_1+1}^{n_1-k_1}(1 - t^j) = 1 - t^{n_1-2k_1+1} + O(t^{n_1-2k_1+2})$ *for* $n_1 - k_1 \geq k_1$ *and zero otherwise. If* $n_1 - 2k_1 + 1 > k_1$, *i.e.* $k_1 < \frac{n_1+1}{3}$, *the probability that the interconnection of minimal systems is reachable and the condition* $\mathrm{rk}(D_1) = k_1$ *is fulfilled is still equal to* $1 - t^{k_1} + O(t^{k_1+1})$. *The same holds for the probability that an interconnection of minimal systems with* $\mathrm{rk}(D_1) = k_1$ *is reachable. But to compute these probabilities for* $\frac{n_1+1}{3} \leq k_1 \leq \frac{n_1}{2}$, *it would be necessary to investigate how the condition* $\mathrm{rk}(D_1) = k_1$ *influences the condition that* P_1 *and* Q_2 *have to be left coprime, which seems to be quite difficult.*

4.4.2 Parallel-Series Connection of two Convolutional Codes

In a parallel-series connection the input for the second system consists not only of the output but also of the input of the first system. Hence, one has $k_2 = n_1$ and the interconnection matrices are

$$K = \begin{bmatrix} 0 & 0 \\ I_{k_2-k_1} & 0 \\ 0 & 0 \end{bmatrix}, \quad L = \begin{bmatrix} I_{k_1} \\ 0 \\ I_{k_1} \end{bmatrix}, \quad M = I_{n_2-k_1}.$$

As $\det(I - DK) = 1 \neq 0$, Theorem 59 could be applied. With $Q_2 = \begin{bmatrix} Q_{21} \\ Q_{22} \end{bmatrix}$
where $Q_{21} \in \mathbb{F}[z]^{(n_1-k_1) \times k_2}$ and $Q_{22} \in \mathbb{F}[z]^{k_1 \times k_2}$, the criterion for reachability of the interconnection is that

$$[Q - KP\ L] = \begin{bmatrix} \begin{bmatrix} Q_1 & 0 \\ 0 & Q_{21} \\ 0 & Q_{22} \end{bmatrix} - \begin{bmatrix} 0 & 0 \\ P_1 & 0 \\ 0 & 0 \end{bmatrix} \begin{bmatrix} I \\ 0 \\ I \end{bmatrix} \end{bmatrix} = \begin{bmatrix} Q_1 & 0 & I \\ -P_1 & Q_{21} & 0 \\ 0 & Q_{22} & I \end{bmatrix}$$

has to be left prime, which is true if and only if $\begin{bmatrix} -P_1 & Q_{21} \\ -Q_1 & Q_{22} \end{bmatrix}$ is left prime.

Theorem 63.
Let $(A_i, B_i, C_i, D_i) \in \mathbb{F}^{\delta_i \times \delta_i} \times \mathbb{F}^{\delta_i \times k_i} \times \mathbb{F}^{(n_i-k_i) \times \delta_i} \times \mathbb{F}^{(n_i-k_i) \times k_i}$ be minimal representations of the non-catastrophic convolutional codes $\mathfrak{C}(A_i, B_i, C_i, D_i)$ for $i = 1, 2$. Then, the probability that $(\mathcal{A}, \mathcal{B}, \mathcal{C}, \mathcal{D})$ is a minimal representation for the parallel-series connected code $\mathfrak{C}(\mathcal{A}, \mathcal{B}, \mathcal{C}, \mathcal{D})$ is equal to

$$1 - t^{k_1} + O\left(t^{k_1+1}\right).$$

Proof.
According to the proof of Lemma 13, one has to show that the probability that
$\begin{bmatrix} -P_1 & Q_{21} \\ -Q_1 & Q_{22} \end{bmatrix}$ is left prime under the condition that $\begin{bmatrix} Q_i \\ P_i \end{bmatrix}$ are right prime for
$i = 1, 2$, is equal to $1 - t^{k_1} + O(t^{k_1+1})$.
One proceeds as in the proof for a series interconnection of linear systems and therefore, defines $A_i := \{(P_i, Q_i)$ not right coprime$\}$ for $i = 1, 2$,
$A_3 := \{\begin{bmatrix} -P_1 & Q_{21} \\ -Q_1 & Q_{22} \end{bmatrix}$ not left prime$\}$ and $A_{ij} := A_i \cap A_j$. Then, the sought-after probability is equal to

$$1 - \frac{\Pr(A_3) - \Pr(A_{13}) - \Pr(A_{23}) + \Pr(A_{123})}{1 - \Pr(A_1) - \Pr(A_2) + \Pr(A_{12})}.$$

At first, $\Pr(A_3)$ should be computed, which is equal to the probability that $\begin{bmatrix} Q_1 & Q_{22} \\ P_1 & Q_{21} \end{bmatrix}$
is not left prime since changing the sign of columns and interchanging rows does not affect the property to be left prime.
One defines $\tilde{Q}_2 := \begin{bmatrix} Q_{22} \\ Q_{21} \end{bmatrix}$ and considers

$$G := \begin{bmatrix} Q_1 & \tilde{Q}_2^H \\ P_1 & \end{bmatrix} := \begin{bmatrix} Q_1 & \tilde{Q}_2 \\ P_1 & \end{bmatrix} \begin{bmatrix} I & 0 \\ 0 & U_2 \end{bmatrix},$$

where $\tilde{Q}_2^H$ is in Hermite form and $U := \begin{bmatrix} I & 0 \\ 0 & U_2 \end{bmatrix}$ is unimodular. Furthermore, one
defines $Q_1^H := Q_1 U_1$ to be the Hermite form of Q_1.

Choose $z_* \in \overline{\mathbb{F}}$ such that $G(z_*)$ is not of full row rank. At first, one wants to eliminate the entries of Q_1 that are fixed (zeros or ones) due to degree constraints. Therefore, consider the rows whose diagonal elements are constant equal to 1; denote the indices of these rows by $i_1, \ldots, i_s$ with $0 \leq s \leq k_1 - 1$ (since $\deg(\det(Q_1)) = \delta_1 \geq 1$). It follows that the other elements of Q_1 that are in one of these rows are identically zero. The same holds for the entries of the columns $i_1, \ldots, i_s$ that are above the diagonal. By adding appropriate multiples of rows $i_1, \ldots, i_s$ of $G(z_*)$ to rows further down, one could achieve that all entries in the columns $i_1, \ldots, i_s$ but the ones on the diagonal are nullified. Thus, one could delete the rows and columns with indices $i_1, \ldots, i_s$ without changing the property that the matrix has no full row rank at z_*. Denote the matrix that is achieved by this row operations and deletion process by $\hat{G}$. The remaining entries of Q_1 and P_1 in $\hat{G}$ are unchanged as well as the entries of $\tilde{Q}_2^H$ that are on the diagonal or above. Moreover, in $\tilde{Q}_2^H$, only rows have been deleted (while in P_1, only columns have been deleted), which affects that the remaining part of it is no longer lower triangular but has a echelon form. Hence, the number of fixed zeros at the end of a row of $\hat{G}$ is strictly decreasing with the index of the row and the last entry of each row that is no fixed zero is a (unchanged) diagonal element of the matrix $\tilde{Q}_2^H$. Furthermore, the difference of a entry of $\tilde{Q}_2^H$ (below the diagonal) in G and the corresponding entry in $\hat{G}$ is a sum of products of entries that were deleted when constructing $\hat{G}$ out of G. Consequently, if we fix the deleted entries, a entry of $\hat{G}$ is fixed at z_* if and only the corresponding entry of G is fixed at z_*.

Therefore, it is possible to treat $\hat{G}(z_*)$ with the same iteration procedure as the matrix $[P_1 \ Q_2^H]$ in the proof for a series interconnection of linear systems. Here, the remaining parts of Q_1 and P_1 play the role of P_1 there, which is possible since all fixed zeros or ones (due to degree conditions) in Q_1 have been deleted, and the remaining part of $\tilde{Q}_2^H$ plays the role of Q_2^H there. Hereby, the echelon form ensures that adding multiples of a row to rows further down does not change the last entry of a row that is no fixed zero, i.e. the diagonal entries of $\tilde{Q}_2^H$ that are contained in $\hat{G}$ remain unchanged. Assume that the iteration stops in the row which had index i in the original matrix G and index $\hat{i}$ in $\hat{G}$. Then, one has $k_1 + i - 1 - s - (\hat{i} - 1)$ polynomials (from Q_1 or $\tilde{Q}_2^H$ for $i \leq k_1$ and from P_1 or $\tilde{Q}_2^H$ for $i > k_1$) that are fixed at z_* by the other entries of G. Moreover, the ii-entry of $\tilde{Q}_2^H$ is fixed to zero, which additionally, leads to the factor $t^{n_1 - i}$ according to (2.1) (see also proof of Theorem 19). Consequently, the probability is $O(t^{k_1 - s + i - \hat{i} + n_1 - i}) = O(t^{k_1 + 1})$ if $n_1 - s - \hat{i} \geq 1$. $\hat{G}$ has $n_1 - s$ rows. Hence, if the iteration stops before the last row of $\hat{G}$, one has $\hat{i} < n_1 - s$ and therefore, the probability is $O(t^{k_1 + 1})$. If $\hat{i} = n_1 - s$, i.e. the iteration continues till the last row, one has $k_1 + n_1 - s - (\hat{i} - 1) = k_1 + 1$ polynomials that are fixed at z_*, which gives a factor for the probability of at least t^{k_1}. Hence, if Q_1^H or $\tilde{Q}_2^H$ have no simple form, which contributes a factor that is $O(t)$, the probability is $O(t^{k_1 + 1})$.

If Q_1^H and $\tilde{Q}_2^H$ are of simple form, G is of the form $\begin{bmatrix} Q_1 & I_{k_2 - 1} & & 0 \\ P_1 & \tilde{q}_{k_2, 1} & \cdots & \tilde{q}_{k_2, k_2} \end{bmatrix}$. As above, we eliminate the fixed zeros and ones (due to degree restrictions) in Q_1 by constructing the matrix $\hat{G}$. Again, denote the indices of the deleted rows in G by $i_1, \ldots, i_s$. Furthermore, set $m_j := j + |\{i_l \leq j, \ l = 1, \ldots, s\}|$ for $j = 1, \ldots, k_1 + k_2 - s$

to achieve $\hat{g}_{ij} = g_{m_i,m_j}$ for $i = 1, \ldots, k_2 - s$ and $j = 1, \ldots, k_1 + k_2 - s$. One gets that $G(z_*)$ is not of full row rank if and only if $\hat{G}(z_*)$ is not of full row rank. This is true if and only if $p^{(1)}_{n_1-k_1,m_i}(z_*) = \hat{g}_{k_2-s,i}(z_*) = \sum_{j=1}^{k_2-1-s} \hat{g}_{ji}\tilde{q}_{k_2,m_j}(z_*)$ for $i \in \{1, \ldots, k_1 - s\}$, $\hat{g}_{k_2-s,i+k_1-s}(z_*) = \sum_{j=1}^{k_2-1-s} \hat{g}_{j,i+k_1-s}\tilde{q}_{k_2,m_j}(z_*)$ for $i \in \{i_1, \ldots, i_s\}$ and $\tilde{q}_{k_2,k_2}(z_*) = 0$. Since the difference of $\hat{g}_{k_2-s,i+k_1-s}(z_*)$ and $\tilde{q}_{k_2,i}(z_*)$ is a sum of products of the values at z_* of entries that were deleted when constructing $\hat{G}$ out of G, the $k_1 + 1$ polynomials $p^{(1)}_{n_1-k_1,i}$ for $i \in \{1, \ldots, k_1\} \setminus \{i_1, \ldots, i_s\}$, $\tilde{q}_{k_2,i}$ for $i \in \{i_1, \ldots, i_s\}$ and $\tilde{q}_{k_2,k_2}$ are fixed at z_* by the values of the other entries of $G(z_*)$. Proceeding as in the part of the proof for Theorem 12 concerning simple form (see also proof for series connection of two systems), one gets that the probability for these conditions is equal to $t^{k_1} + O(t^{k_1+1})$ and hence, $\Pr(A_3) = t^{k_1} + O(t^{k_1+1})$, too. It is already known that $\Pr(A_2) = O(t^{k_2}) = O(t^{k_1+1})$. Thus, it only remains to compute $\Pr(A_{13})$. As in the case of series connection, one has only to consider simple form. Moreover, it is known that the corresponding matrices lie in A_1 if and only if there exists $\tilde{z}_* \in \overline{\mathbb{F}}$ with $q^{H_1}_{k_1,k_1}(\tilde{z}_*) = \sum_{l=1}^{k_1} p^{(1)}_{1l} u^{(1)}_{l,k_1}(\tilde{z}_*) = \cdots = \sum_{l=1}^{m_1} p^{(1)}_{n_1-k_1,l} u^{(1)}_{l,k_1}(\tilde{z}_*) = 0$. Again, as in the proof for series connection, there exists l_0 such that $u_{l_0,k_1}(\tilde{z}_*) \neq 0$ and $q_{l_0,l_0} \neq 1$. Therefore, $l_0 \notin \{i_1, \ldots, i_s\}$ and the doubly fixed polynomial $p^{(1)}_{n_1-k_1,l_0}$ has an upper bound for its degree of at least 1, which ensures that Lemma 7 (c) could be applied to it. Still following the lines of the proof for series connection, one gets $\Pr(A_{13}) = O(t^{k_1+1})$ and finally, $1 - t^{k_1} + O(t^{k_1+1})$ for the desired probability. $\square$

As in the previous subsection, one has $M = I$ and therefore, the observability of (A_i, B_i, C_i, D_i) for $i = 1, 2$ implies the observability of $(\mathcal{A}, \mathcal{B}, \mathcal{C}, \mathcal{D})$. If one wants to show this with the help of Theorem 10, one needs the right primeness of

$$\begin{bmatrix} Q - KP \\ MP \end{bmatrix} = \begin{bmatrix} Q_1 & 0 \\ -P_1 & Q_{21} \\ 0 & Q_{22} \\ P_1 & 0 \\ 0 & P_2 \end{bmatrix}, \text{ which is given since } \begin{bmatrix} Q_1 & 0 \\ P_1 & 0 \\ 0 & Q_2 \\ 0 & P_2 \end{bmatrix} \text{ is right prime.}$$

Thus, one has:

Theorem 64.
Let $(A_i, B_i, C_i, D_i) \in \mathbb{F}^{\delta_i \times \delta_i} \times \mathbb{F}^{\delta_i \times k_i} \times \mathbb{F}^{(n_i-k_i) \times \delta_i} \times \mathbb{F}^{(n_i-k_i) \times k_i}$ be minimal representations of the non-catastrophic convolutional codes $\mathfrak{C}(A_i, B_i, C_i, D_i)$ for $i = 1, 2$. Then, the probability that $(\mathcal{A}, \mathcal{B}, \mathcal{C}, \mathcal{D})$ is a minimal representation for the parallel-series connected code $\mathfrak{C}(\mathcal{A}, \mathcal{B}, \mathcal{C}, \mathcal{D})$ and that $\mathfrak{C}(\mathcal{A}, \mathcal{B}, \mathcal{C}, \mathcal{D})$ is non-catastrophic is equal to

$$1 - t^{k_1} + O\left(t^{k_1+1}\right).$$

Furthermore, the probability that this interconnection is minimal is equal to the probability that the single systems are minimal and $\begin{bmatrix} -P_1 & Q_{21} \\ -Q_1 & Q_{22} \end{bmatrix}$ is left prime.
Hence, analogous to the previous subsection, one has the following theorem.

Theorem 65.
Let $(A_i, B_i, C_i, D_i) \in \mathbb{F}^{\delta_i \times \delta_i} \times \mathbb{F}^{\delta_i \times k_i} \times \mathbb{F}^{(n_i - k_i) \times \delta_i} \times \mathbb{F}^{(n_i - k_i) \times k_i}$ be randomly for $i = 1, 2$. Then, the probability that (A, B, C, D) is a minimal representation for the parallel-series connected code $\mathfrak{C}(A, B, C, D)$ and that $\mathfrak{C}(A, B, C, D)$ is non-catastrophic is equal to

$$\prod_{i=1}^{2} (1 - t^{k_i} - t^{n_i - k_i} + O(t^{\min(k_i, n_i - k_i) + 1})) \cdot (1 - t^{k_1} + O(t^{k_1 + 1})) =$$

$$= 1 - 2t^{k_1} - t^{k_2} - t^{n_1 - k_1} - t^{n_2 - k_2} + O(t^{\min(k_1, k_2, n_1 - k_1, n_2 - k_2) + 1}).$$

4.4.3 Interleaved Parallel-Series Connection

The most frequently occuring type of series connection in the coding literature (see e.g., [3], [12]) is given by interconnection matrices of the following form:

$$K = \begin{bmatrix} 0 & 0 \\ \pi_1 & 0 \\ 0 & 0 \end{bmatrix}, \quad L = \begin{bmatrix} I_{k_1} \\ 0 \\ \pi_2 \end{bmatrix}, \quad M = [0 \; I_{n_2 - k_2}]$$

with permutation matrices $\pi_1 \in S_{k_2 - k_1}$ and $\pi_2 \in S_{k_1}$. We call this a interleaved parallel-series connection. The difference to a parallel-series connection is that the second part of the input vector for the second system is not necessarily identical with the input vector for the first system but its entries are a permutation of the entries of the input vector for the first system. Again, it holds $\det(I - DK) = 1 \neq 0$ and Theorem 10 could be applied. One gets the criterion that the interconnection is reachable if and only if $\begin{bmatrix} \pi_1 P_1 \\ \pi_2 Q_1 \end{bmatrix} Q_2$ is left prime. If P_1 and Q_1 are randomly, the probability for that is not affected by the values for π_1 and π_2 and hence, one has:

Theorem 66.
Let $(A_i, B_i, C_i, D_i) \in \mathbb{F}^{\delta_i \times \delta_i} \times \mathbb{F}^{\delta_i \times k_i} \times \mathbb{F}^{(n_i - k_i) \times \delta_i} \times \mathbb{F}^{(n_i - k_i) \times k_i}$ be minimal representations of the non-catastrophic convolutional codes $\mathfrak{C}(A_i, B_i, C_i, D_i)$ for $i = 1, 2$. Then, the probability that (A, B, C, D) is a minimal representation for the interleaved parallel-series connected code $\mathfrak{C}(A, B, C, D)$ is equal to

$$1 - t^{k_1} + O\left(t^{k_1 + 1}\right).$$

The most frequently occuring type of parallel connection is also an interleaved one (see e.g., [2], [41]). In the coding literatur, such interleaved parallel connected codes are called turbo codes and will be considered in the following subsection.

4.4.4 Turbo Codes

For this type of network of convolutional codes, the interconnection matrices are

$$K = 0, \quad L = \begin{bmatrix} \pi_1 \\ \vdots \\ \pi_N \end{bmatrix}, \quad M = I$$

with $\pi_i \in S_k$ for $i = 1, \ldots, N$. Again $\det(I - DK) = 1 \neq 0$ and Theorem 10 implies that this interconnection is reachable if and only if

$$[Q(z) - KP(z)\,L] = \begin{bmatrix} Q_1(z) & & & \pi_1 \\ & \ddots & & \vdots \\ & & Q_N(z) & \pi_N \end{bmatrix}$$

is left prime. Subtracting $\pi_i \cdot \pi_1^{-1}$ times the first block of rows from the i-th block for $i = 2, \ldots, N$, transforms this matrix to $\begin{bmatrix} Q_1 & & & \pi_1 \\ -\pi_2\pi_1^{-1}Q_1 & Q_2 & & 0 \\ \vdots & & \ddots & \vdots \\ -\pi_N\pi_1^{-1}Q_1 & & & Q_N & 0 \end{bmatrix}$, which

is left prime if and only if the following chain of matrices is left prime:

$$\begin{bmatrix} -\pi_2\pi_1^{-1}Q_1 & Q_2 & & 0 \\ \vdots & & \ddots & \\ -\pi_N\pi_1^{-1}Q_1 & & & Q_N \end{bmatrix} \rightsquigarrow \begin{bmatrix} -\pi_2\pi_1^{-1}Q_1 & Q_2 & & 0 \\ 0 & -\pi_3\pi_2^{-1}Q_2 & & \\ \vdots & \vdots & & \ddots \\ 0 & -\pi_N\pi_2^{-1}Q_2 & & Q_N \end{bmatrix}$$

$$\rightsquigarrow \begin{bmatrix} -\pi_2\pi_1^{-1}Q_1 & Q_2 & & 0 \\ & \ddots & & \ddots \\ 0 & & -\pi_N\pi_{N-1}^{-1}Q_{N-1} & Q_N \end{bmatrix}$$

$$\rightsquigarrow \begin{bmatrix} -\pi_1^{-1}Q_1 & \pi_2^{-1}Q_2 & & 0 \\ & \ddots & & \ddots \\ 0 & & -\pi_{N-1}^{-1}Q_{N-1} & \pi_N^{-1}Q_N \end{bmatrix}.$$

This is true if and only if $\pi_1^{-1}Q_1, \ldots, \pi_N^{-1}Q_N$ are mutually left coprime.

Remark 28.
  (i) *Very often turbo codes are used which consist of an interconnection of identical codes, i.e. one has $Q_1 = \cdots = Q_N$. In this case, it is necessary that the matrices π_i are pairwise different, which is not possible for $k! < N$, i.e. in particular, in the case of single-input, where $k = 1$.*
  (ii) *If the polynomial matrices Q_i are arbitrary, the probability that $\pi_1^{-1}Q_1, \ldots \pi_N^{-1}Q_N$ are mutually left coprime is equal to the probability that $Q_1, \ldots, Q_N$ are mutually left coprime.*

If one wants to compute the probability that a interconnected system representing a turbo code is reachable, it is easier not to apply Theorem 10 with K and L as defined at the beginning of this subsection but to trace this problem back to the computation of the probability of reachability for a parallel connection with $K = 0$ and $L = [I \ \cdots \ I]^\top$. Doing this, it does not matter that M is different for turbo codes and parallel connected systems because this matrix has no affect on the reachability. Therefore, neglecting the output of the interconnected system, a turbo code where the constituent codes are represented by (A_i, B_i, C_i, D_i) for $i = 1, \ldots, N$ could be regarded as identical with the code represented by the parallel connection of $(A_i, B_i\pi_i, C_i, D_i\pi_i)$. Since for fixed matrices π_i and randomly chosen (A_i, B_i, C_i, D_i), the matrix-quadruple $(A_i, B_i\pi_i, C_i, D_i\pi_i)$ reaches every element of $\mathbb{F}^{\delta_i \times \delta_i} \times \mathbb{F}^{\delta_i \times k_i} \times \mathbb{F}^{(n_i-k_i)\times\delta_i} \times \mathbb{F}^{(n_i-k_i)\times k_i}$ the same number of times, the probability that the parallel connection of $(A_i, B_i\pi_i, C_i, D_i\pi_i)$ for $i = 1, \ldots, N$ is reachable is equal to the probability that the parallel connection of (A_i, B_i, C_i, D_i) for $i = 1, \ldots, N$ is reachable. Therefore, using Theorem 31, one gets the following theorem:

Theorem 67.
Let $(A_i, B_i, C_i, D_i) \in \mathbb{F}^{\delta_i \times \delta_i} \times \mathbb{F}^{\delta_i \times k} \times \mathbb{F}^{(n_i-k)\times\delta_i} \times \mathbb{F}^{(n_i-k)\times k}$ be randomly for $i = 1, \ldots, N$. Then, the probability that $(\mathcal{A}, \mathcal{B}, \mathcal{C}, \mathcal{D})$ is a minimal representation for the turbo code $\mathfrak{C}(\mathcal{A}, \mathcal{B}, \mathcal{C}, \mathcal{D})$ is equal to

$$1 - \sum_{y=1}^{k+1} \binom{N}{y} t^k + O(t^{k+1}).$$

Remark 29.
Viewing a turbo code as a code represented by the parallel connection of the linear systems $(A_i, B_i\pi_i, C_i, D_i\pi_i) = (A, B\pi_i, C, D\pi_i)$ for $i = 1, \ldots, N$ also makes it possible to improve the statement of Remark 28 (i). By the Hautus test (see Theorem 2), the parallel connected system is reachable if and only if the matrix
$$\begin{bmatrix} zI - A & & 0 & B\pi_1 \\ & \ddots & & \vdots \\ 0 & & zI - A & B\pi_N \end{bmatrix}$$
has full row rank for each $z \in \overline{\mathbb{F}}$. If z is an eigenvalue of A, the column rank of the first $N\delta$ columns of this matrix is at most $N(\delta - 1)$.

Therefore, for left primeness of the whole matrix it is necessary that $\begin{bmatrix} B\pi_1 \\ \vdots \\ B\pi_N \end{bmatrix}$ *has (at least) N linearly independent columns, which is only possible if $k \geq N$. This clearly is a stronger condition that $k! \geq N$ from Remark 28 (i).*

Next, we consider what happens if the node system are not chosen completely randomly but only amongst minimal systems, i.e. we require the constituent codes to be non-catastrophic. One gets that the probability for the interconnection to be a minimal representation of the corresponding convolutional code is asymptotically equal to the probability of mutual left coprimeness.

Theorem 68.
Let $(A_i, B_i, C_i, D_i) \in \mathbb{F}^{\delta_i \times \delta_i} \times \mathbb{F}^{\delta_i \times k_i} \times \mathbb{F}^{(n_i-k_i)\times\delta_i} \times \mathbb{F}^{(n_i-k_i)\times k_i}$ be minimal representations of the non-catastrophic convolutional codes $\mathfrak{C}(A_i, B_i, C_i, D_i)$ for $i = 1, \ldots, N$. Then, the probability that $(\mathcal{A}, \mathcal{B}, \mathcal{C}, \mathcal{D})$ is a minimal representation for the turbo code $\mathfrak{C}(\mathcal{A}, \mathcal{B}, \mathcal{C}, \mathcal{D})$ is equal to

$$1 - \sum_{y=2}^{k+1} \binom{N}{y} t^k + O(t^{k+1}).$$

Proof.
One has to compute the probability that $\pi_1^{-1}Q_1, \ldots \pi_N^{-1}Q_N$ are mutually left coprime under the condition that (A_i, B_i, C_i, D_i) are minimal for $i = 1, \ldots, N$. To do this, one proceeds as in the beginning of Section 3.3 and defines $E_i := \{(P_i, Q_i) \text{ not right coprime}\}$. Moreover, set
$E_{N+1} := \{\pi_1^{-1}Q_1, \ldots \pi_N^{-1}Q_N \text{ not mutually left coprime}\}$ and $E_{ij} := E_i \cap E_j$. Analogous to (3.5), one gets that the sought-after probability is equal to

$$\frac{1 - \Pr(\cup_{i=1}^{N+1} E_i)}{1 - \Pr(\cup_{i=1}^{N} E_i)} = \frac{1 - \sum_{\emptyset \neq I \subset \{1,\ldots,N+1\}} (-1)^{|I|-1} \Pr(E_I)}{1 - \sum_{\emptyset \neq I \subset \{1,\ldots,N\}} (-1)^{|I|-1} \Pr(E_I)} =$$

$$= 1 - \frac{\Pr(E_{N+1}) + O(\sum_{i=1}^{N} \Pr(E_{i,N+1}))}{1 - \sum_{\emptyset \neq I \subset \{1,\ldots,N\}} (-1)^{|I|-1} \Pr(E_I)}. \tag{4.3}$$

It remains to show that $\Pr(E_{i,N+1}) = t \cdot O(\Pr(E_{N+1}))$ for $i = 1, \ldots, N$. Then, the theorem follows from Theorem 22 because

$$\begin{bmatrix} -\pi_1^{-1}Q_1 & \pi_2^{-1}Q_2 & & 0 \\ & \ddots & \ddots & \\ 0 & & -\pi_{N-1}^{-1}Q_{N-1} & \pi_N^{-1}Q_N \end{bmatrix} \cdot \begin{bmatrix} U_1 & & 0 \\ & \ddots & \\ 0 & & U_N \end{bmatrix} =$$

$$= \begin{bmatrix} -\pi_1^{-1}Q_1 U_1 & \pi_2^{-1}Q_2 U_2 & & 0 \\ & \ddots & \ddots & \\ 0 & & -\pi_{N-1}^{-1}Q_{N-1}U_{N-1} & \pi_N^{-1}Q_N U_N \end{bmatrix}$$

and hence, for the computation of $\Pr(E_{N+1})$, one could assume that $\pi_i^{-1}Q_i$ are in Hermite form for $i = 1, \ldots, N$, which implies $\Pr(E_{N+1}) = P_m(N)$.
Since P_i and Q_i are right coprime if and only if $\pi_i^{-1}P_i$ and $\pi_i^{-1}Q_i$ are right coprime and $\deg_j(P_i) \leq \deg_j(Q_i)$ for $j = 1, \ldots, k$ if and only if $\deg_j(\pi_i^{-1}P_i) \leq \deg_j(\pi_i^{-1}Q_i)$ for $j = 1, \ldots, k$, one could define $\tilde{Q}_i := \pi_i^{-1}Q_i$ and $\tilde{P}_i := \pi_i^{-1}P_i$ for $i = 1, \ldots, N$ and compute the probability that $\tilde{Q}_1, \ldots, \tilde{Q}_N$ are not mutually left coprime and $\tilde{Q}_i$ and $\tilde{P}_i$ are not right coprime. Moreover, as frequently done before, one could assume $\tilde{Q}_1, \ldots, \tilde{Q}_N$ to be in simple form (see proof of Theorem 21). According to the proof of Theorem 12, in this case, $\tilde{Q}_i$ and $\tilde{P}_i$ are not right prime if and only if there exists $z_0 \in \bar{\mathbb{F}}$ such that

$$\tilde{q}_{kk}^{H_i}(z_0) = \sum_{l=1}^{k} \tilde{p}_{1l}^{(i)} u_{lk}(z_0) = \cdots = \sum_{l=1}^{k} \tilde{p}_{n_i-k,l}^{(i)} u_{lk}(z_0) = 0. \tag{4.4}$$

If one fixes $\tilde{Q}_1, \ldots, \tilde{Q}_N$ not mutually left coprime (which fixes U, too), one has only finitely many possibilities for z_0 to be a zero of $\tilde{q}_{kk}^{H_i}$. For each of these possibilities, it was shown in the proof of Theorem 12 that there exists $l_0 \in \{1, \ldots, k\}$ such that $u_{l_0,k}(z_0) \neq 0$. Hence, it follows from Lemma 7 (b) that the probability that P_i fulfils (4.4) is at most $t^{n_i-k} \leq t$. Consequently, $\Pr(E_{i,N+1}) = t \cdot O(\Pr(E_{N+1}))$ and one is done. $\qquad\square$

Since $M = I$ for turbo codes, one could employ the same reasoning as for series connections of convolutional codes and additionally, gets the following two theorems:

Theorem 69.
Let $(A_i, B_i, C_i, D_i) \in \mathbb{F}^{\delta_i \times \delta_i} \times \mathbb{F}^{\delta_i \times k} \times \mathbb{F}^{(n_i-k) \times \delta_i} \times \mathbb{F}^{(n_i-k) \times k}$ be minimal representations of the non-catastrophic convolutional codes $\mathfrak{C}(A_i, B_i, C_i, D_i)$ for $i = 1, \ldots, N$. Then, the probability that $(\mathcal{A}, \mathcal{B}, \mathcal{C}, \mathcal{D})$ is a minimal representation for the turbo code $\mathfrak{C}(\mathcal{A}, \mathcal{B}, \mathcal{C}, \mathcal{D})$ and that $\mathfrak{C}(\mathcal{A}, \mathcal{B}, \mathcal{C}, \mathcal{D})$ is non-catastrophic is

$$1 - \sum_{y=2}^{k+1} \binom{N}{y} t^k + O(t^{k+1}).$$

Theorem 70.
Let $(A_i, B_i, C_i, D_i) \in \mathbb{F}^{\delta_i \times \delta_i} \times \mathbb{F}^{\delta_i \times k} \times \mathbb{F}^{(n_i-k) \times \delta_i} \times \mathbb{F}^{(n_i-k) \times k}$ be randomly for $i = 1, \ldots, N$. Then, the probability that $(\mathcal{A}, \mathcal{B}, \mathcal{C}, \mathcal{D})$ is a minimal representation for the turbo code $\mathfrak{C}(\mathcal{A}, \mathcal{B}, \mathcal{C}, \mathcal{D})$ and that $\mathfrak{C}(\mathcal{A}, \mathcal{B}, \mathcal{C}, \mathcal{D})$ is non-catastrophic is

$$\left(1 - \sum_{y=2}^{k+1} \binom{N}{y} t^k + O(t^{k+1})\right) \prod_{i=1}^{N} (1 - t^k - t^{n_i-k} + O(t^{\min(k, n_i-k)})).$$

Proof.
This theorem follows from the preceding theorem, Remark 4 and Theorem 27. $\quad\square$

Remark 30.
In [9], it is shown that for a turbo code with two constituent codes, $\mathrm{rk}(D_1) = \mathrm{rk}(D_2) = k$ is a sufficient condition to lower bound the free distance of the concatenated code by the maximum of the free distances of the constituent codes plus 1. According to Lemma 4, the probability that this condition is fulfilled is equal to $\prod_{j=n_1-2k+1}^{n_1-k}(1 - t^j) \cdot \prod_{j=n_2-2k+1}^{n_2-k}(1 - t^j) = 1 - t^{n_1-2k+1} - t^{n_2-2k+1} + O(t^{\min(n_1,n_2)-2k+2})$ if $n_i - k \geq k$ for $i = 1, 2$ and zero otherwise.
In contrast to Remark 27, where we made similar considerations for series interconnection, the matrices D_1 and D_2 have no influence on the reachability and observability of the single systems and the interconnected system here. Consequently, if one wants to calculate a lower bound for the probability that the interconnected code is non-catastrophic and its free distance can be lower bounded as above, one could just multiply the probabilities for these two properties.

4.5 Interconnection of Convolutional and Block Codes

In this section, we consider two types of interconnected codes, which were introduced in [8]. Both are variations of a series connection, where the first code is a block code and the second code is a convolutional code. We will call the first model a series connection and the second model a truncated series connection.

4.5.1 Series Connection

Here, a sequence of codewords of a block code with generator matrix $G \in \mathbb{F}^{m \times k}$ of full column rank and where $m > k$ serves as input for a linear system $(A, B, C, D) \in \mathbb{F}^{\delta \times \delta} \times \mathbb{F}^{\delta \times m} \times \mathbb{F}^{(n-k) \times \delta} \times \mathbb{F}^{(n-k) \times m}$ defining a convolutional code of rate $\frac{m}{n-k+m}$ and degree δ. The final codeword consists of the output of (A, B, C, D) and the input sequence for the block code. Hence, (A, BG, C, DG) is a representation of the interconnected code, which is a convolutional code of rate $\frac{k}{n}$ and degree δ. If one fixes G with full column rank and chooses B randomly, BG reaches every element of $\mathbb{F}^{\delta \times k}$ exactly $t^{-\delta(m-k)}$ times. Since every element occurs the same number of times, it follows that the probability that (A, BG) is reachable coincides with the probability that $(A, \tilde{B})$ with $\tilde{B} \in \mathbb{F}^{\delta \times k}$ is reachable. Analogously, the probability that (A, BG, C) is minimal coincides with the probability that $(A, \tilde{B}, C)$ is minimal. Because these probabilities are independent of G, they stay the same if one also chooses G randomly and does not fix it any more. Therefore, one has the following theorem:

Theorem 71.
Let $G \in \mathbb{F}^{m \times k}$ of full column rank and $(A, B, C, D) \in \mathbb{F}^{\delta \times \delta} \times \mathbb{F}^{\delta \times m} \times \mathbb{F}^{(n-k) \times \delta} \times \mathbb{F}^{(n-k) \times m}$ be randomly. Then, it holds:

(i) *The probability that (A, BG, C, DG) is a minimal representation of the series connected code $\mathfrak{C}(A, BG, C, DG)$ is equal to*

$$\prod_{j=k}^{\delta+k-1} (1 - t^j).$$

(ii) *The probability that (A, BG, C, DG) is a minimal representation of the series connected code $\mathfrak{C} = \mathfrak{C}(A, BG, C, DG)$ and that $\mathfrak{C}$ is non-catastrophic is*

$$1 - t^k - t^{n-k} + O(t^{\min(k, n-k)+1}).$$

Proof.

(i) The statement follows from the preceding considerations and Theorem 26.

(ii) See the preceding considerations and Theorem 27. □

Next, we want to consider what happens if (A, B, C, D) is not chosen completely randomly but one has the restriction that this matrix quadruple is a minimal representation of the second code of the series connection.

Theorem 72.
Let $G \in \mathbb{F}^{m \times k}$ of full column rank and $(A, B, C, D) \in \mathbb{F}^{\delta \times \delta} \times \mathbb{F}^{\delta \times m} \times \mathbb{F}^{(n-k) \times \delta} \times \mathbb{F}^{(n-k) \times m}$ be randomly. Then, it holds:

(i) *The probability that (A, BG, C, DG) is a minimal representation of the series connected code $\mathfrak{C}(A, BG, C, DG)$ if (A, B, C, D) is a minimal representation of $\mathfrak{C}(A, B, C, D)$ is equal to*

$$\prod_{j=k}^{\delta+k-1} (1 - t^j) \cdot \left(\prod_{j=m}^{\delta+m-1} (1 - t^j) \right)^{-1} = \frac{1 - t^k + O(t^{k+1})}{1 - t^m + O(t^{m+1})} = 1 - t^k + O(t^{k+1}).$$

(ii) The probability that (A, BG, C, DG) is a minimal representation of the series connected code $\mathfrak{C} = \mathfrak{C}(A, BG, C, DG)$ and that $\mathfrak{C}$ is non-catastrophic if (A, B, C, D) is a minimal representation of the non-catastrophic code $\mathfrak{C}(A, B, C, D)$ is

$$\frac{1 - t^k - t^{n-k} + O(t^{\min(k,n-k)+1})}{1 - t^m - t^{n-k} + O(t^{\min(m,n-k)+1})} = \begin{cases} 1 - t^k + O(t^{k+1}) & \text{for } n - k \geq k \\ 1 + O(t^{n-k+1}) & \text{for } n - k < k \end{cases}.$$

Proof.

Since the surjectivity of $\mathcal{R}(A, BG) = \mathcal{R}(A, B) \cdot \begin{pmatrix} G & & 0 \\ & \ddots & \\ 0 & & G \end{pmatrix}$ implies the surjectivity of $\mathcal{R}(A, B)$, the probability that (A, BG) and (A, B) are reachable is equal to the probability that (A, BG) is reachable. Analogously, the probability that (A, BG, C) and (A, B, C) are minimal is equal to the probability that (A, BG, C) is minimal. Using the formula for conditional probability as well as the preceding results, one gets the statements of the theorem. $\qquad\square$

Remark 31.
If one interprets the input sequence for the block code as coefficients of a polynomial, one could regard this interconnection as the series connection of two convolutional codes, where the first one has degree 0.

The main reason why we got interested in this type of interconnected code is that in [8] the authors provide a sufficient condition for such a concatenation to be of maximum distance profile if the corresponding parameters fulfill certain criteria. Therefore, in the following, let $k \geq 2$, $n = k + 1$ and $\delta = 1$, i.e. $(A, B, C, D) \in \mathbb{F} \times \mathbb{F}^{1 \times m} \times \mathbb{F} \times \mathbb{F}^{1 \times m}$. For this case, one has the following theorem:

Theorem 73. *[8, Theorem 25]*
If $k \geq 2$ and the matrix $\begin{bmatrix} I_k \\ DG \\ CBG \end{bmatrix} \in \mathbb{F}^{(k+2) \times k}$ is the generator matrix of a MDS block code, then the series connected code $\mathfrak{C}(A, BG, C, DG)$ has rate $\frac{k}{k+1}$, degree 1 and is a maximum distance profile convolutional code.

Therefore, we want to compute the probability that a $[k + 2, k]$-block code is MDS.

Theorem 74.
The probability that a $[k + 2, k]$-block code is MDS is equal to

$$(1 - t)^{k+1} \prod_{i=2}^{k} (1 - it) = 1 - \sum_{i=2}^{k+1} i \cdot t + O(t^2) = 1 - \frac{(k+3)k}{2} \cdot t + O(t^2).$$

Proof.
According to Theorem 50, the needed probability is equal to the probability that every square submatrix of a random matrix $P \in \mathbb{F}^{2 \times k}$ is nonsingular. The formula for this

probability should be shown per induction with respect to k. For $k = 1$, the two entries of P have to be unequal to zero, i.e. the probability is equal to $(1-t)^2$. Now, consider

$P = \begin{bmatrix} p_{11} & \cdots & p_{1k} \\ p_{21} & \cdots & p_{2k} \end{bmatrix}$. Per induction, the probability that every square submatrix

of $\begin{bmatrix} p_{11} & \cdots & p_{1,k-1} \\ p_{21} & \cdots & p_{2,k-1} \end{bmatrix}$ is nonsingular is equal to $(1-t)^k \prod_{i=2}^{k-1}(1-it)$. Additionally,

one has the conditions $p_{1k} \neq 0$, $p_{2k} \neq 0$ and $p_{1i}p_{2k} - p_{2i}p_{1k} \neq 0$ for $i = 1, \ldots, k-1$. Hence, if all entries of P but p_{2k} are fixed (unequal to zero), one has to exclude (besides zero) the values $\frac{p_{2i}p_{1k}}{p_{1i}}$ for $i = 1, \ldots, k-1$ for p_{2k}. These values are pairwise different since otherwise, there would exist $1 \leq i \neq j \leq k-1$ with $\frac{p_{2i}}{p_{1i}} = \frac{p_{2j}}{p_{1j}}$. But this would imply that the square submatrix consisting of the i-th and j-th column was singular. Thus, one has $1 - t$ possibilities for p_{1k} and $1 - kt$ possibilities for p_{2k}. The statement follows since $(1-t)^k \prod_{i=2}^{k-1}(1-it)(1-t)(1-kt) = (1-t)^{k+1} \prod_{i=2}^{k}(1-it)$. Note that this probability is equal to zero if $t^{-1} \leq k$, which corresponds to the fact that you cannot exclude k values for an entry in this case. □

Remark 32.

(i) *The condition of Theorem 73 implies that $BG \neq 0_{1 \times k}$, i.e. that (A, BG) is reachable, as well as that $C \neq 0$, i.e. that (A, C) is observable and therefore, the concatenated convolutional code is of degree 1 and non-catastrophic.*

(ii) *Since Theorem 73 only considers matrices P of a special form, the preceding theorem is not sufficient to get a formula for the probability that the series connected code is of maximum distance profile but some additional investigations are necessary.*

(iii) *Since Theorem 73 only provides a sufficient condition for the interconnected code to be of maximum distance profile, one only could achieve a lower bound for the probability that it has this property.*

Theorem 75.
For randomly chosen matrices $B \in \mathbb{F}^{1 \times m}$, $C \in \mathbb{F}$, $D \in \mathbb{F}^{1 \times m}$ and $G \in \mathbb{F}^{m \times k}$ such that G has full column rank, the probability that every square submatrix of $\begin{bmatrix} DG \\ CBG \end{bmatrix}$ is

nonsingular is equal to $(1-t)^{k+2} \prod_{i=2}^{k}(1-it)$.
Consequently, the probability that the series connection of a randomly chosen $[m, k]$-block code with $k \geq 2$ and a convolutional code $\mathfrak{C}(A, B, C, D)$ with randomly chosen $(A, B, C, D) \in \mathbb{F} \times \mathbb{F}^{1 \times m} \times \mathbb{F} \times \mathbb{F}^{1 \times m}$ is of maximum distance profile is at least

$$(1-t)^{k+2} \prod_{i=2}^{k}(1-it) = 1 - \sum_{i=1}^{k+1} i \cdot t + O(t^2) = 1 - \frac{(k+2)(k+1)}{2} \cdot t + O(t^2).$$

Proof.

For $C = 0$, $\begin{bmatrix} DG \\ CBG \end{bmatrix}$ has a entry equal to zero and hence, a singular square submatrix.

Since G is of full column rank, the matrix $\begin{bmatrix} DG \\ BG \end{bmatrix}$ reaches every element of $\mathbb{F}^{2 \times k}$

exactly $t^{-2(m-k)}$ times, if $D, B \in \mathbb{F}^{1 \times m}$ are chosen arbitrarily. Thus, for $C \neq 0$, the matrix $\begin{bmatrix} DG \\ CBG \end{bmatrix}$ reaches every element of $\mathbb{F}^{2 \times k}$ exactly $(1 - t)t^{-2(m-k)}$ times if D, C and B are chosen randomly. In particular, every element of $\mathbb{F}^{2 \times k}$ is obtained with the same probability if (A, B, C, D) is chosen randomly with $C \neq 0$. Consequently, the result follows from the preceding theorem, where the additional factor $1 - t$ arises from the condition $C \neq 0$. □

Combining the preceding theorem with Remark 23, leads to the following result:

Corollary 19.
The probability that the series connection of a randomly chosen $[m, k]$-block code with $k \geq 2$ and a convolutional code $\mathfrak{C}(A, B, C, D)$ with randomly chosen $(A, B, C, D) \in \mathbb{F} \times \mathbb{F}^{1 \times m} \times \mathbb{F} \times \mathbb{F}^{1 \times m}$ is strongly MDS is at least

$$(1 - t)^{k+2} \prod_{i=2}^{k} (1 - it).$$

For the case $k = 1$ (and $n = 2$, $\delta = 1$), one could use the following theorem:

Theorem 76. *[8, Theorem 31]*
If for each $v \in \{D, CB, CAB, CBGCB - CABGD\}$, it holds $vG \neq 0$, then the series connected code $\mathfrak{C}(A, BG, C, DG)$, has rate $\frac{1}{2}$, degree 1 and is a maximum distance profile convolutional code.

Computing the probability that the conditions of this theorem are fulfilled, one gets a lower bound for the probability that the interconnection is of maximum distance profile in the case $k = 1$:

Theorem 77.
The probability that the series connection of a randomly chosen $[m, 1]$-block code and a convolutional code $\mathfrak{C}(A, B, C, D)$ with randomly chosen $(A, B, C, D) \in \mathbb{F} \times \mathbb{F}^{1 \times m} \times \mathbb{F} \times \mathbb{F}^{1 \times m}$ is of maximum distance profile is at least

$$(1 - t)^3 (1 - 2t).$$

Proof.
Let $G \in \mathbb{F}^m \setminus \{0\}$ be the generator matrix of the block code. It holds $vG \neq 0$ for each $v \in \{D, CB, CAB\}$ if and only if $A \neq 0$, $C \neq 0$, $BG \neq 0$ and $DG \neq 0$. Assuming these conditions, $(CBGCB - CABGD)G = 0$ if and only if $(B - \frac{A}{C}D)G \neq 0$. Thus, one has $(t^{-1} - 1)^2$ possibilities for the choice of A and C, and for fixed G, $t^{-m} - t^{-(m-1)}$ possibilities for D. If A, C and D are fixed with these conditions, one additionally has $BG \notin \{0, \frac{A}{C}DG\}$. Since $\frac{A}{C}DG \neq 0$, this leads to $t^{-m} - 2t^{-(m-1)}$ possibilities for B. Hence, the total probability is equal to
$$\frac{(t^{-1} - 1)^2 \cdot (t^{-m} - t^{-(m-1)}) \cdot (t^{-m} - 2t^{-(m-1)})}{t^{-2-2m}} = (1 - t)^3 (1 - 2t). \qquad \square$$

As above, combining the previous result with Remark 23 provides another statement:

Corollary 20.
The probability that the series connection of a randomly chosen $[m, 1]$-block code and a convolutional code $\mathfrak{C}(A, B, C, D)$ with randomly chosen $(A, B, C, D) \in \mathbb{F} \times \mathbb{F}^{1 \times m} \times \mathbb{F} \times \mathbb{F}^{1 \times m}$ is strongly MDS is at least

$$(1 - t)^3 (1 - 2t).$$

4.5.2 Truncated Series Connection

Here, one chooses for the first code a generator matrix of the form $G = \begin{pmatrix} I_k \\ P \end{pmatrix} \in \mathbb{F}^{m \times k}$.

For each $u \in \mathbb{F}^k$ which is encoded by this block code, only $Pu \in \mathbb{F}^{m-k}$ is transmitted to the second code, which is a convolutional code of rate $\frac{m-k}{n-2k+m}$ and degree δ defined by a linear system $(A, B, C, D) \in \mathbb{F}^{\delta \times \delta} \times \mathbb{F}^{\delta \times (m-k)} \times \mathbb{F}^{(n-k) \times \delta} \times \mathbb{F}^{(n-k) \times (m-k)}$. The final codeword consists of the output of (A, B, C, D) and the input sequence for the block code. Hence, (A, BP, C, DP) is a representation for the interconnected code, which is a convolutional code of rate $\frac{k}{n}$ and degree δ. To consider if this representation is a minimal one, set $\tilde{B} = BP$ and $r = \mathrm{rk}(P) \leq \min(m - k, k)$. If $P \in \mathbb{F}^{(m-k) \times k}$ is fixed and $B \in \mathbb{F}^{\delta \times (m-k)}$ is chosen randomly, it holds $\ker(P) \subset \ker(\tilde{B})$ and each element of $\mathbb{F}^{\delta \times k}$ with this property is reached by $\tilde{B}$ exactly $t^{-\delta(k-r)}$ times. Since for $T \in Gl_k(\mathbb{F})$, $\mathcal{R}(A, \tilde{B}) \cdot T$ is of full row rank if and only if $\mathcal{R}(A, \tilde{B})$ is of full row rank, one could assume that $\tilde{B} = (\tilde{b}_1, \ldots, \tilde{b}_r, 0, \ldots, 0)$, where $\tilde{b}_i \in \mathbb{F}^\delta$ for $i = 1, \ldots, k$ and $(\tilde{b}_1, \ldots, \tilde{b}_r)$ reaches every element of $\mathbb{F}^{\delta \times r}$ exactly $t^{-\delta(k-r)}$ times. It follows $\mathrm{rk}(\mathcal{R}(A, \tilde{B})) = \mathrm{rk}(\mathcal{R}(A, (\tilde{b}_1, \ldots, \tilde{b}_r)))$ and hence, according to Theorem 26, the probability that $(A, \tilde{B})$ is reachable is equal to $\prod_{j=r}^{\delta+r-1}(1 - t^j)$. This formula does not depend on the concrete choice of P as long as $\mathrm{rk}(P) = r$. Finally, one gets the following theorem:

Theorem 78.
Let $P \in \mathbb{F}^{(m-k) \times k}$ and $(A, B, C, D) \in \mathbb{F}^{\delta \times \delta} \times \mathbb{F}^{\delta \times (m-k)} \times \mathbb{F}^{(n-k) \times \delta} \times \mathbb{F}^{(n-k) \times (m-k)}$ be chosen randomly. Then, it holds:

(i) *The probability that (A, BP, C, DP) is a minimal representation of the truncated series connected code $\mathfrak{C}(A, BP, C, DP)$ is equal to*

$$\sum_{r=1}^{\min(m-k,k)} t^{(m-k)k} \cdot N(m - k, k, r) \cdot \prod_{j=r}^{\delta+r-1} (1 - t^j).$$

(ii) *The probability that (A, BP, C, DP) is a minimal representation of the truncated series connected code $\mathfrak{C} = \mathfrak{C}(A, BP, C, DP)$ and that $\mathfrak{C}$ is non-catastrophic is*

$$\sum_{r=1}^{\min(m-k,k)} t^{(m-k)k} \cdot N(m - k, k, r) \cdot (1 - t^r - t^{n-k} + O(t^{\min(r,n-k)+1})).$$

Moreover, analogously to Theorem 72, one has:

Theorem 79.
*Let $P \in \mathbb{F}^{(m-k)\times k}$ and $(A, B, C, D) \in \mathbb{F}^{\delta\times\delta} \times \mathbb{F}^{\delta\times(m-k)} \times \mathbb{F}^{(n-k)\times\delta} \times \mathbb{F}^{(n-k)\times(m-k)}$
be chosen randomly. Then, it holds:*

(i) *The probability that (A, BP, C, DP) is a minimal representation of the truncated
series connected code $\mathfrak{C}(A, BP, C, DP)$ if (A, B, C, D) is a minimal representation
of $\mathfrak{C}(A, B, C, D)$ is equal to*

$$\sum_{r=1}^{\min(m-k,k)} t^{(m-k)k} \cdot N(m-k, k, r) \prod_{j=r}^{\delta+r-1} (1 - t^j) \cdot \left(\prod_{j=m-k}^{\delta+m-k-1} (1 - t^j) \right)^{-1}.$$

(ii) *The probability that (A, BP, C, DP) is a minimal representation of the truncated
series connected code $\mathfrak{C} = \mathfrak{C}(A, BP, C, DP)$ and that $\mathfrak{C}$ is non-catastrophic if
(A, B, C, D) is a minimal representation of the non-catastrophic code $\mathfrak{C}(A, B, C, D)$
is*

$$\frac{\sum_{r=1}^{\min(m-k,k)} t^{(m-k)k} \cdot N(m-k, k, r) \cdot (1 - t^r - t^{n-k} + O(t^{\min(r,n-k)+1}))}{1 - t^{m-k} - t^{n-k} + O(t^{\min(m-k,n-k)+1})}.$$

Again, we proceed considering the case $k \geq 2$, $n = k+1$ and $\delta = 1$ since for
truncated series connection, one has the following analogon to Theorem 73:

Theorem 80. *[8, Theorem 29]*
*If $k \geq 2$ and the matrix $\begin{bmatrix} I_k \\ DP \\ CBP \end{bmatrix} \in \mathbb{F}^{(k+2)\times k}$ is the generator matrix of a MDS block
code, then the truncated series connected code $\mathfrak{C}(A, BP, C, DP)$, which has degree 1 and
rate $\frac{k}{k+1}$, is a maximum distance profile convolutional code.*

Computing the probability that the condition of the previous theorem is fulfilled,
leads to:

Theorem 81.
*For $k \geq 2$ and randomly chosen matrices $B \in \mathbb{F}^{1\times(m-k)}$, $C \in \mathbb{F}$, $D \in \mathbb{F}^{1\times(m-k)}$ and
$P \in \mathbb{F}^{(m-k)\times k}$, the probability that every square submatrix of $\begin{bmatrix} DP \\ CBP \end{bmatrix}$ is nonsingular
is equal to*

$$(1 - t)^3 \cdot (1 - t^{m-k}) \cdot (1 - t^{m-k-1}) \cdot \prod_{j=2}^{k} (1 - (j+1) \cdot t + j \cdot t^2).$$

*Consequently, the probability that the truncated series connection of a block code with
generator matrix $G = \begin{pmatrix} I_k \\ P \end{pmatrix} \in \mathbb{F}^{m\times k}$, where P is chosen randomly, and a convolutional*

code $\mathfrak{C}(A, B, C, D)$ with randomly chosen $(A, B, C, D) \in \mathbb{F} \times \mathbb{F}^{1 \times (m-k)} \times \mathbb{F} \times \mathbb{F}^{1 \times (m-k)}$ is of maximum distance profile is at least

$$(1 - t)^3 \cdot (1 - t^{m-k}) \cdot (1 - t^{m-k-1}) \cdot \prod_{j=2}^{k} \left(1 - (j+1) \cdot t + j \cdot t^2\right).$$

Proof.

For $C = 0$, $\begin{bmatrix} DP \\ CBP \end{bmatrix}$ has a entry equal to zero and hence, a singular square sub-matrix. Moreover, it does not influence the probability which nonzero value is taken by C. Therefore, the overall probability is equal to $1 - t$ times the probability that every square submatrix of $\begin{bmatrix} DP \\ BP \end{bmatrix}$ is nonsingular. To compute this proba-bility, set $l := m - k$ and write $P = (p_1 \ldots p_k)$ with $p_i \in \mathbb{F}^l$ for $i = 1, \ldots, k$, i.e. $\begin{bmatrix} DP \\ BP \end{bmatrix} = \begin{bmatrix} Dp_1 & \cdots & Dp_k \\ Bp_1 & \cdots & Dp_k \end{bmatrix}$. Since, in particular, it is necessary that $(Dp_1) \cdot (Bp_2) - (Bp_1)(Dp_2) \neq 0$, it follows that B and D have to be linearly in-dependent. Otherwise, there would exist $\lambda \in \mathbb{F}$ with $D = \lambda B$, which implies $(Dp_1) \cdot (Bp_2) - (Bp_1)(Dp_2) = \lambda(Bp_1) \cdot (Bp_2) - \lambda(Bp_1)(Bp_2) = 0$.

Using these notations and considerations, in the following, we will prove the stated formula per induction with respect to k and start with $k = 2$. In this case, one has to look at $\begin{bmatrix} Dp_1 & Dp_2 \\ Bp_1 & Dp_2 \end{bmatrix} \in \mathbb{F}^{2 \times 2}$. As shown above for D and B, p_1 and p_2 have to be linearly independent, too. But if p_1 and p_2 are linearly independent, one is in the situation of Theorem 75. Consequently, one has to multiply the probability that two elements from $\mathbb{F}^l$ are linearly independent with the formula from Theo-rem 75 for the case $k = 2$, in which the factor $1 - t$ for the condition $C \neq 0$ is already included. In summary, one gets $(1 - t^l) \cdot (1 - t^{l-1}) \cdot (1 - t)^4 \cdot (1 - 2t) = (1 - t)^3 \cdot (1 - t^{m-2}) \cdot (1 - t^{m-3}) \cdot (1 - 3t + 2t^2)$.

For the step from $k - 1$ to k, one uses that every square summatrix of $\begin{bmatrix} DP \\ BP \end{bmatrix} = \begin{bmatrix} Dp_1 & \cdots & Dp_k \\ Bp_1 & \cdots & Dp_k \end{bmatrix}$ is nonsingular if and only if every square submatrix of $M := \begin{bmatrix} Dp_1 & \cdots & Dp_{k-1} \\ Bp_1 & \cdots & Dp_{k-1} \end{bmatrix}$ is nonsingular and $Dp_k \neq 0$, $Bp_k \neq 0$ and $(Bp_iD - Dp_iB)p_k \neq 0$ for $i = 1, \ldots, k - 1$. Per induction, one knows that the probability that every square submatrix of M is nonsingular is equal to $(1 - t)^3 \cdot (1 - t^l) \cdot (1 - t^{l-1}) \cdot \prod_{j=2}^{k-1} \left(1 - (j+1) \cdot t + j \cdot t^2\right)$. We fix D, B and $p_1, \ldots, p_{k-1}$ with this property. To determine the additional factor for the overall probability due to the conditions on p_k, define

$$A_i := \{(Bp_iD - Dp_iB)p_k = 0\} \quad \text{for} \quad i = 1, \ldots, k - 1$$
$$A_k := \{Dp_k = 0\}, \quad A_{k+1} := \{Bp_k = 0\}.$$

Then, according to the inclusion-exclusion principle (see Lemma 2), the sought-after additional factor is equal to

$$1 - \sum_{\emptyset \neq I \subset \{1,\dots,k+1\}} (-1)^{|I|-1} \Pr(A_I) \quad \text{with } A_I = \bigcap_{i \in I} A_i.$$

Next, we show $A_I = A_k \cap A_{k+1}$ for every $I \subset \{1,\dots,k+1\}$ with $|I| = 2$, which implies $A_I = A_k \cap A_{k+1}$ for $|I| \geq 2$. First consider $A_k \cap A_i$ with $i \in \{1,\dots,k-1\}$. In this case, one has $Dp_k = 0$ as well as $(Bp_i D - Dp_i B)p_k = 0$, i.e. $Dp_i Bp_k = 0$, which implics $Bp_k = 0$ since $Dp_i \neq 0$ because it is a (one-dimensional) square submatrix of M. Hence, $A_k \cap A_i \subset A_k \cap A_{k+1}$. But the other implication $A_k \cap A_{k+1} \subset A_k \cap A_i$ is obviously true and therefore, $A_k \cap A_i = A_k \cap A_{k+1}$. The proof of $A_{k+1} \cap A_i = A_k \cap A_{k+1}$ for $i = 1,\dots,k-1$ could be done completely analogue. It remains to show $A_i \cap A_j \subset A_k \cap A_{k+1}$ for $i,j \in \{1,\dots,k-1\}$ with $i \neq j$. To this end, one firstly shows $A_i \cap A_j \cap A_k^C = \emptyset$, where A_k^C denotes the complementary set of A_k. In A_k^C it holds $Dp_k \neq 0$ and thus, $(Bp_i D - Dp_i B)p_k = 0$ is equivalent to $Bp_i = \frac{Dp_i Bp_k}{Dp_k}$. Analogously, $(Bp_j D - Dp_j B)p_k = 0$ is equivalent to $Bp_j = \frac{Dp_j Bp_k}{Dp_k}$. But then, $Bp_i Dp_j - Dp_i Bp_j = \frac{Dp_i Bp_k}{Dp_k} Dp_j - Dp_i \frac{Dp_j Bp_k}{Dp_k} = 0$, a contradiction to the fact that the square submatrix of M formed by the i-th and j-th column is nonsingular. Consequently, $A_i \cap A_j \cap A_k^C = \emptyset$. In the same way, one could proof $A_i \cap A_j \cap A_{k+1}^C = \emptyset$, which shows $A_i \cap A_j \subset A_k \cap A_{k+1}$.

Thus, the additional factor for the probability due to the conditions on p_k simplifies to

$$1 - \sum_{i=1}^{k+1} \Pr(A_i) + \sum_{j=2}^{k+1} \binom{k+1}{j} (-1)^j \Pr(A_k \cap A_{k+1}) =$$

$$= 1 - \sum_{i=1}^{k+1} \Pr(A_i) + k \cdot \Pr(A_k \cap A_{k+1}).$$

Here, for the last step, the identity $\sum_{j=0}^{k+1} (-1)^j \binom{k+1}{j} = 0$ was used. To compute $\Pr(A_i)$ note that $Bp_i D - Dp_i B \neq 0 \in \mathbb{F}^{1 \times l}$ since otherwise $Bp_i Dp_j - Dp_i Bp_j = 0$ for $j = 1,\dots,k-1$ with $j \neq i$, which is in contradiction to the assumptions on M (as we already saw). Similarly, B and D have a nonzero entry. Therefore, $\Pr(A_i) = t$ for $i = 1,\dots,k+1$. Since we saw at the beginning of the proof that the condition on M implies that D and B are linearly independent, one additionally gets $\Pr(A_k \cap A_{k+1}) = t^2$, i.e.

$$1 - \sum_{i=1}^{k+1} \Pr(A_i) + k \cdot \Pr(A_k \cap A_{k+1}) = 1 - (k+1) \cdot t + k \cdot t^2.$$

Multiplying this factor with the probability that every square submatrix of M is nonsingular, completes the proof of the whole theorem. □

Again, using Remark 23 enables us to get a further result.

Corollary 21.
The probability that the truncated series connection of a randomly chosen $[m, k]$-block code with $k \geq 2$ and a convolutional code $\mathfrak{C}(A, B, C, D)$ with randomly chosen $(A, B, C, D) \in \mathbb{F} \times \mathbb{F}^{1 \times m} \times \mathbb{F} \times \mathbb{F}^{1 \times m}$ is strongly MDS is at least

$$(1 - t)^3 \cdot (1 - t^{m-k}) \cdot (1 - t^{m-k-1}) \cdot \prod_{j=2}^{k} \left(1 - (j+1) \cdot t + j \cdot t^2 \right).$$

As in the previous subsection, we finalize with considering the case $k = 1$ (and $n = 2$, $\delta = 1$), where one could use the following theorem, which is analogue to Theorem 76:

Theorem 82. *[8, Theorem 33]*
If for each $v \in \{D, CB, CAB, CBPCB - CABPD\}$, it holds $vP \neq 0$, then the truncated series connected code $\mathfrak{C}(A, BP, C, DP)$, which has degree 1 and rate $\frac{1}{2}$, is a maximum distance profile convolutional code.

Using this, one gets the following theorem:

Theorem 83.
The probability that the truncated series connection of a randomly chosen $[m, 1]$-block code and a convolutional code $\mathfrak{C}(A, B, C, D)$ with randomly chosen $(A, B, C, D) \in \mathbb{F} \times \mathbb{F}^{1 \times m} \times \mathbb{F} \times \mathbb{F}^{1 \times m}$ is of maximum distance profile is at least

$$(1 - t)^4 (1 - 2t).$$

Proof.
The criterion provided by the preceding theorem could only be fulfilled if P is not the zero vector. However, this implies that P is of full (column) rank and one is in the situation of Theorem 77. Therefore, one just has to multiply the formula from there with the probability that P is unequal to the zero vector, which is $1 - t$. ☐

Finally, Remark 23 leads to another corollary:

Corollary 22.
The probability that the truncated series connection of a randomly chosen $[m, 1]$-block code and a convolutional code $\mathfrak{C}(A, B, C, D)$ with randomly chosen $(A, B, C, D) \in \mathbb{F} \times \mathbb{F}^{1 \times m} \times \mathbb{F} \times \mathbb{F}^{1 \times m}$ is strongly MDS is at least

$$(1 - t)^4 (1 - 2t).$$

4.6 Probability of MDP Convolutional Codes

In the preceding section, we computed, in particular, the probability that special types of series connection of convolutional code and block code have the MDP property. In this section, we will look on general convolutional codes and compute the probabiliy that such a code is of maximum distance profile. In the first subsection, this is done for the case that the rate is $1/n$ and the degree δ is 1, while the second subsection deals with the general case, effecting that no exact probability but a upper and a lower bound are obtained.

4.6.1 Probability of MDP Codes with Rate $1/n$ and Degree 1

Since the generator matrix consists only of one column, one could write it as $G(z) = \sum_{i=0}^{\delta} g_i z^i = g_0 + g_1 z$ with $g_0, g_1 \in \mathbb{F}^n$. Moreover, $L = \lfloor \frac{\delta}{k} \rfloor + \lfloor \frac{\delta}{n-k} \rfloor = 1 + \lfloor \frac{1}{n-1} \rfloor$. Hence, $L = 1$ for $n \geq 3$ and $L = 2$ for $n = 2$.
We will start with the case $n = 2$, i.e. $n = k + 1$ as at the end of the two preceding subsections.

Theorem 84.
The probability that a randomly chosen convolutional code with parameters $k = 1$, $n = 2$ and $\delta = 1$ is of maximum distance profile is $\frac{(1-t)^2(1-2t)}{1+t} = 1 - 5t + O(t^2)$.
In particular, there exists no binary MDP convolutional code with these parameters.

Proof.
According to Theorem 53, one has to compute the probability that each full size minor

$$
\text{of } \mathcal{G}_2 = \begin{bmatrix} g_{0,1} & 0 & 0 \\ g_{0,2} & 0 & 0 \\ g_{1,1} & g_{0,1} & 0 \\ g_{1,2} & g_{0,2} & 0 \\ 0 & g_{1,1} & g_{0,1} \\ 0 & g_{1,2} & g_{0,2} \end{bmatrix} \text{ that is not trivially zero is nonzero, under the condition}
$$

that $g_1 \neq 0$ to ensure that $\delta = 1$. This is true if and only if $0 \notin \{g_{0,1}, g_{0,2}, g_{1,1}, g_{1,2}\}$ and $g_{1,1}g_{0,2} - g_{1,2}g_{0,1} \neq 0$. The probability that these conditions are fulfilled is equal to $(1 - t)^3(1 - 2t)$. Dividing by the probability for the condition $g_1 \neq 0$, which is $1 - t^2$, one gets the stated result. $\qquad\square$

Remark 33.
Without the condition $g_1 \neq 0$, one gets the same probability as the lower bound of Theorem 77. The constraint $g_1 \neq 0$, which ensures that the degree is 1, corresponds to the property of (A, BG) from Theorem 77 being reachable, which was not taken as a condition there. Thus, for comparison, we should take the probability value $(1 - t)^3(1 - 2t)$. Since it coincides with the lower bound from Theorem 77, a series concatenation of a block and a convolutional code has at least the probability of being MDP than a general convolutional code with the same parameters.

Finally, the case $n \geq 3$ should be considered.

Theorem 85.
The probability that a randomly chosen convolutional code with parameters $k = 1$, $n \geq 3$ and $\delta = 1$ is of maximum distance profile is

$$(1 - t^n)^{-1}(1 - t)^{n+1} \prod_{i=2}^{n-1}(1 - it) = 1 - \frac{n(n+1)}{2} \cdot t + O(t^2).$$

Proof.
In this case, one has to consider the fullsize minors of $\mathcal{G}_1 = \begin{bmatrix} g_0 & 0 \\ g_1 & g_0 \end{bmatrix}$ under the condition $g_1 \neq 0$. All not trivially zero fullsize minors of $\mathcal{G}_1$ are nonzero if and only if $g_{0,i} \neq 0$ for $i = 1, \ldots, n$ and $g_{1,j} \neq \frac{g_{1,i} \cdot g_{0,j}}{g_{0,i}}$ for $i < j$ and $j = 2, \ldots, n$. This is fulfilled with probability $(1 - t)^n \prod_{j=2}^{n}(1 - (j-1)t) = (1 - t)^{n+1} \prod_{j=2}^{n-1}(1 - jt)$. Finally, one has to divide by $1 - t^n$, which is the probability that $g_1 \neq 0$. $\qquad\square$

4.6.2 Probability Bounds for General MDP Convolutional Codes

In [25], it has been shown that the property of a convolutional code to be of maximum distance profile is generic. In other words, for each choice for the parameters of a convolutional code, there exists an extension field of $\mathbb{F}$ such that the probability that a convolutional code over this extension field and with these parameters is of maximum distance profile is unequal to zero. In the following, a lower and an upper bound for this probability should be obtained.

Theorem 86.
The probability that a convolutional code with fixed parameters n, k and δ has maximum distance profile is upper bounded by

$$(1 - t)^{(n-k)k} = 1 - (n - k)k \cdot t + O(t^2).$$

Proof.
For each convolutional code $\mathfrak{C}$, choose an arbitrary representation (A, B, C, D) with $\mathfrak{C} = \mathfrak{C}(A, B, C, D)$. For being MDP it is necessary that all entries of D are nonzero, see Theorem 57. Hence, the statement follows from the fact that $D \in \mathbb{F}^{(n-k) \times k}$. $\qquad\square$

Theorem 87.
Let $\mathbb{F}$ be finite with cardinality $|\mathbb{F}| = t^{-1}$. There exists $d \in \mathbb{N}$ such that the probability that a convolutional code over the extension field $\mathbb{F}^d$ has maximum distance profile is lower bounded by

$$1 - \frac{C(n, k, \delta) \cdot t^d}{\prod_{j=\delta}^{\delta+k-1}(1 - t^{jd})} = 1 - C(n, k, \delta) \cdot t^d + O(t^{2d}).$$

with

$$C(n,k,\delta) = \sum_{s=1}^{(L+1)\min(k,n-k)} s(L+1) \sum_{\{i_1,...,i_s\}\subset\{1,...,(L+1)(n-k)\}} \prod_{x=1}^{s} \max(0, \lceil \frac{i_x}{n-k} \rceil \cdot k - (x-1))$$

$$\leq \sum_{s=1}^{(L+1)\min(k,n-k)} s(L+1) \binom{(L+1)(n-k)}{s} \binom{(L+1)k}{s}.$$

Proof.
For each convolutional code $\mathfrak{C}$, choose $(A,B,C,D) \in \mathbb{F}^{\delta\times\delta} \times \mathbb{F}^{\delta\times k} \times \mathbb{F}^{(n-k)\times\delta} \times \mathbb{F}^{(n-k)\times k}$ with $\mathfrak{C} = \mathfrak{C}(A,B,C,D)$. In particular, (A,B) is reachable. Since all minimal representations of a convolutional code are conjugated (see Theorem 56), they lead to the same matrix $\mathcal{T}_L$. Moreover, each convolutional code with the same parameters has the same number of minimal representations, namely $|Gl_\delta(\mathbb{F})|$. Thus, the probability that a convolutional code is not MDP is equal to the probability that there exists a not trivially zero minor of $\mathcal{T}_L$ that is zero, under the condition that (A,B) is reachable. This probability is upper bounded by the quotient of the probability of such a zero minor divided by the probability of reachability.
Each minor of $\mathcal{T}_L$ is a polynomial in the entries of A, B, C and D, whose degree is upper bounded by $L+1$. The corresponding code is not MDP if and only if the product f of all polynomials corresponding to not trivially zero minors is zero. According to [25], there exists an extension field $\mathbb{F}^d$ over which there exists a convolutional code with the given parameters n,k,δ and therefore, f is not the zero polynomial over this extension field. Hence, one could apply the Schwartz-Zippel Lemma (see Lemma 3) and gets that the probability that f is zero is upper bounded by $\deg(f) \cdot t$. Consequently, it remains to show that $C(n,k,\delta)$ is an upper bound for the degree of f.
According to Definition 1.3 of [11], the not trivially zero minors could be described in the following way: For a $s \times s$-minor, choose a set of rows $\{i_1,\ldots,i_s\}$ and a corresponding set $\{j_1,\ldots,j_s\}$ of columns fulfilling $j_x \leq \lceil \frac{i_x}{n-k} \rceil \cdot k$ for $x = 1,\ldots s$. Thus, if the rows $\{i_1,\ldots,i_s\}$ are fixed, one has $\lceil \frac{i_1}{n-k} \rceil \cdot k$ possibilities for the first column. Since all s columns have to be chosen differently, one has $\lceil \frac{i_x}{n-k} \rceil \cdot k - (x-1)$ possibilities for the x-th column. Because this number might be negative, one has to take the maximum of this number and 0 in the above formula. Finally, the factor $s(L+1)$ arises from the fact that the degree of each $s \times s$-minor is upper bounded by $(L+1)s$ as the degree of each entry is upper bounded by $L+1$.
If one considers the product of all minors of $\mathcal{T}_L$ (not only those which are not trivially zero), one gets the weaker bound stated in the last line of the theorem. □

Corollary 23.
The probability that a convolutional code with fixed parameters is of maximum distance profile is $1 + O(t)$ for $t \to 0$.

4.7 Probability of MDS Block Codes

In Theorem 74, we computed the probability that a $[k+2, k]$-block code is MDS to be able to obtain probability bounds for interconnections of block and convolutional codes. But the probability that a block code is MDS is of interest on its own and therefore, we want to consider it for arbitrary parameters now. To this end, we will use Theorem 50, i.e. we have to compute the probability that every minor of the non-systematic part of the generator matrix, which is an element of $\mathbb{F}^{(n-k)\times k}$, is nonzero. However, we will only achieve bounds for this probability.

Theorem 88.
The probability that a random $[n, k]$-block code is MDS is upper bounded by $(1-t)^{(n-k)k}$.

Proof.
It clearly is necessary that all entries of the non-systematic part of the generator matrix are non-zero and $(1-t)^{(n-k)k}$ is the probability that this is fulfilled. □

Theorem 89.
If $1/t \geq n$, the probability that a $[n, k]$-block code is MDS is lower bounded by $1 - \sum_{j=1}^{\min(k,n-k)} \binom{k}{j}\binom{n-k}{j} \cdot t$.

Proof.
It is known that if $1/t \geq n$, there exists a $[n, k]$ MDS block code over $\mathbb{F}$. Therefore, one could apply the Schwartz-Zippel Lemma (Lemma 3) to the product of all the minors of the non-systematic part of the generator matrix of the code since it is not the zero polynomial when we view it as a polynomial with the entries of the non-systematic part as variables. The bound follows because the degree of this polynomial is equal to $\sum_{j=1}^{\min(k,n-k)} \binom{k}{j}\binom{n-k}{j}$. □

Corollary 24.
The probability that a $[n, k]$-block code is MDS is $1 + O(t)$ for $t \to 0$.

4.8 Random Linear Network Coding

We start this section with a short summary about network codes; see [23] for more details. One considers a network represented by a graph $\Gamma = (\mathcal{V}, \mathcal{E})$ with set of vertices $\mathcal{V}$ and set of directed edges $\mathcal{E}$. The set of vertices is partitioned into three disjunct subsets $\mathcal{V}_1, \mathcal{V}_2$ and $\mathcal{V}_3$, where the elements of $\mathcal{V}_1$ have no ingoing edges and are called source nodes and the elements of $\mathcal{V}_3$ have no outgoing edges and are called sink nodes. Moreover, we assume that there are no edges originating at a source node and terminating at a sink node. Additionally, set $r := |\mathcal{V}_1|$, $l := |\mathcal{V}_2|$, $d := |\mathcal{V}_3|$ and associate with the corresponding vertices the values $u_1, \ldots, u_r, x_1, \ldots, x_l \in \mathbb{F}$, $y_1, \ldots, y_d \in \mathbb{F}^r$ with $t^{-1} = |\mathbb{F}| = 2^m$ for some $m \in \mathbb{N}$. The values of the source nodes represent the coding message. Furthermore, the value of a vertex is formed as a linear combination of the values of the origins of its ingoing edges. The graph Γ defines a so-called network coding problem, which is said to be solvable if the coefficients for these linear combinations could be chosen in such way that $y_i = (u_1, \ldots, u_r)^\top$ for $i = 1, \ldots, d$. This corresponds to the fact that the original message could be reproduced at each sink node. In the following, we consider two different types of network codes.

4.8.1 Delay-free Acyclic Network Coding

In this subsection, we assume that the network contains no cycles - in other words it is acyclic. This additionally allows us to assume that transmission happens instantaneously and simultaneously, i.e. delay-free. Under these conditions, one gets the following coding equations:

$$x_i = \sum_{j=1}^{l} a_{ij} x_j + \sum_{j=1}^{r} b_{ij} u_j \quad \text{for } i = 1, \ldots, l \tag{4.5}$$

with $A \in \mathbb{F}^{l \times l}$ and $B \in \mathbb{F}^{l \times r}$, where $a_{ij} = 0$ if Γ contains no edge from vertex j of V_2 to vertex i of V_2 and $b_{ij} = 0$ if Γ contains no edge from vertex j of V_1 to vertex i of V_2. Since Γ is acyclic, we could number the vertices in such way that A is lower triangular with zeros on the diagonal. Moreover, the decoding equations are given by

$$y_{i,k} = \sum_{j=1}^{l} c_{kj}^{(i)} x_j \quad \text{for } i = 1, \ldots, d \quad \text{and} \quad k = 1, \ldots, r \tag{4.6}$$

with $C^{(i)} \in \mathbb{F}^{r \times l}$ for $i = 1, \ldots, d$. As above, $C^{(i)}$ contains fixed zero entries if the corresponding edge is missing in Γ. Combining (4.5) and (4.6), leads to

$$y_i = M_i u \quad \text{for } i = 1, \ldots, d$$

with transfer matrices $M_i := C^{(i)} (I - A)^{-1} B \in \mathbb{F}^{r \times r}$ and $u := (u_1, \ldots, u_r)^\top$. According to the proof of Theorem 2.2. in [23], a network coding problem is solvable if and only if there exist A, B and $C^{(i)}$ for $i = 1, \ldots, d$ with entries from a sufficiently large finite field such that $\det(M_i)$ is nonzero for $i = 1, \ldots, d$.

In the approach of random linear network coding, one chooses the entries of the matrices A and B that are no fixed zeros due to the structure of Γ randomly from the finite field $\mathbb{F}$ and asks for the probability that there exist $C^{(i)}$ for $i = 1, \ldots, d$ that lead to a solution for the network coding problem. Theorem 2.6 of [23] in particular states that this probability is lower bounded by $(1 - dt)^l$ if $t^{-1} > d$ and all entries of A below the diagonal and all entries of B could be chosen randomly, i.e. Γ is the complete graph. Especially, it follows that the probability tends to 1 if t^{-1} - the size of the field $\mathbb{F}$ - tends to infinity. In the following, we set $r = 1$. At first, we give an exact formula for the above probability in this case. Afterwards, we consider the probability that random choice of A, B and $C^{(i)}$ for $i = 1, \ldots, d$ leads to a solution for the network coding problem. For both computations, we will need the following lemma:

Lemma 16. *[23, Lemma 2.3]*
For an acyclic delay-free network, it holds

$$| \det(M_i)| = \left| \det \left(\begin{bmatrix} C^{(i)} & O \\ I - A & B \end{bmatrix} \right) \right| \quad \textit{for } i = 1, \ldots, d.$$

Since we need the probabilty that $\det(M_i) \neq 0$ for $i = 1, \ldots, d$ in some special situations, this lemma will help to prove the following two theorems:

Theorem 90.
Let Γ be acyclic and $r = 1$. If the coding coefficients, i.e. the entries of $B \in \mathbb{F}^l$ as well as the entries below the diagonal of $A \in \mathbb{F}^{l \times l}$, are chosen randomly, the probability that there exist decoding coefficients, i.e. matrices $C^{(i)} \in \mathbb{F}^{1 \times l}$ for $i = 1, \ldots, d$, which lead to a solution of the network coding problem is equal to

$$1 - t^l > (1 - dt)^l.$$

Proof.
We have to compute the probability that there exist $C^{(i)} \in \mathbb{F}^{1 \times l}$ such that

$$\det \left(\begin{bmatrix} C^{(i)} & O \\ I - A & B \end{bmatrix} \right) = \begin{bmatrix} c_1^{(i)} & \cdots & \cdots & c_l^{(i)} & 0 \\ 1 & 0 & \cdots & 0 & b_1 \\ -a_{21} & \ddots & \ddots & \vdots & \vdots \\ \vdots & \ddots & \ddots & 0 & \vdots \\ -a_{ll} & \cdots & -a_{l,l-1} & 1 & b_l \end{bmatrix} \neq 0 \text{ for } i = 1, \ldots, d \text{ if }$$

$B \in \mathbb{F}^l$ and the entries below the diagonal of $A \in \mathbb{F}^{l \times l}$ are chosen randomly. Recall that the other entries of A are fixed zeros.

In the following, we will show per induction with respect to l that the only case in which every choice of $c^{(i)}$ leads to zero determinant is when $b_1 = \cdots = b_l = 0$, which proves the stated formula for the probability.

For $l = 1$, the matrices M_i are of the form $\begin{bmatrix} c^{(i)} & 0 \\ -a & b \end{bmatrix} \in \mathbb{F}^{2 \times 2}$ for $i = 1, \ldots, d$. Thus,

the only case in which every choice of $c^{(i)}$ leads to zero determinant is when $b = 0$.
For $l \geq 2$, expanding $\det(M_i)$ along the 2-th row yields

$$\det(M_i) = \det(M_i^{(l-1)}) \pm b_1 \cdot \det \begin{bmatrix} c_1^{(i)} & \cdots & \cdots & c_l^{(i)} \\ -a_{21} & 1 & & 0 \\ \vdots & \ddots & \ddots & \\ -a_{ll} & \cdots & -a_{l,l-1} & 1 \end{bmatrix} \tag{4.7}$$

where $M_i^{(l-1)}$ is defined to have the same structure as M_i but for a network with $|\mathcal{V}_2| = l - 1$. If $b_1 \neq 0$, expanding the last determinant in (4.7) along the first row yields that one could solve the equation $\det(M_i) = 0$ with respect to $c_1^{(i)}$. Hence, there always is a possibility to choose the matrices $C^{(i)}$ in such way that the matrices M_i are nonsingular. Consequently, it only remains to consider the case $b_1 = 0$. Here, $\det(M_i) = 0$ if and only if $\det(M_i^{(l-1)}) = 0$ and the claim follows per induction.

$\square$

Theorem 91.

If Γ is acyclic, $r = 1$ and the entries of $B \in \mathbb{F}^l$ and $C^{(i)} \in \mathbb{F}^{1 \times l}$ for $i = 1, \ldots, d$ as well as the entries below the diagonal of A are chosen randomly, the probability that one gets a solution for the network coding problem is

$$(1 - t)^d (1 - t^l).$$

Proof.

As in the previous proof (and with the notation from there), one uses (4.7) to show the stated formula per induction with respect to l. For $l = 1$, one gets a solution if and only if $b \neq 0$ and $c^{(i)} \neq 0$ for $i = 1, \ldots, d$, i.e. the probability is equal to $(1 - t)^{d+1}$. For $l \geq 2$, one again distinguishes the cases $b_1 \neq 0$ and $b_1 = 0$. If $b_1 \neq 0$, it follows from (4.7) that for each choice of A, B and $c_2^{(i)}, \ldots, c_l^{(i)}$, there is exactly one value for $c_1^{(i)}$ that yields $\det(M_i) = 0$ and therefore has to be excluded. For $b_1 = 0$, one gets a solution if and only if $\det(M_i^{(l-1)}) \neq 0$, which has a probability of $(1 - t)^d (1 - t^{l-1})$ per induction. Consequently, the overall probability is equal to

$$(1 - t)^{d+1} + t \cdot (1 - t)^d (1 - t^{l-1}) = (1 - t)^d \cdot (1 - t + t - t^l) = (1 - t)^d (1 - t^l).$$

$\square$

Remark 34.

The fact that is possible to achieve a solution for the network coding problem if and only if $B \in \mathbb{F}^l \neq 0$, which was stated in the proof of Theorem 90, could also be seen without considering the matrices M_i. It is obvious that there is no solution if $B = 0$. If $B \neq 0$, assume without restriction that $b_1 \neq 0$ and choose $c_2^{(i)} = \cdots = c_l^{(i)} = 0$ as well as $c_1^{(i)} \neq 0$ to get a solution. Combing Theorems 90 and Theorem 91, makes it possible to achieve a formula for the number of possible solutions, namely $t^{-ld} \cdot (1 - t)^d$. The first factor of this expression gives the total number of matrices $C^{(i)}$ for $i = 1, \ldots, d$ and the second factor is the probability that random choice of these matrices leads to a solution, provided that A and B are chosen in such way that it is possible to solve the network coding problem. This probability is obtained by taking the quotient of the formulas from Theorem 91 and Theorem 90.

Remark 35.

For $r \geq 2$, it clearly is necessary for a solution of the network coding problem that B is of full column rank. This in particular, implies $r \leq l$, which corresponds to the fact that r parts of information cannot be forwarded by less than r network nodes. Thus, the probability te get a solution is upper bounded by the probability that B is of full column rank, which is equal to $\prod_{j=l-r+1}^{l}(1 - t^j)$. In the case $r = l$, one has $\det(M_i) = \det(C^{(i)}) \cdot \det(B)$ for $i = 1, \ldots, d$, and therefore, the probability that it is possible to get a solution is equal to $t^{r^2} \cdot |Gl_r(\mathbb{F})| = \prod_{j=1}^{r}(1 - t^j)$. Moreover, the probability that one gets a solution if the matrices $C^{(i)}$ are chosen randomly, too, is upper bounded by $\left(\prod_{j=l-r+1}^{l}(1 - t^j)\right)^{d+1}$ and for $r = l$, it is equal to $\left(\prod_{j=1}^{r}(1 - t^j)\right)^{d+1}$.

4.8.2 Memory-free Convolutional Network Coding

So far, we only considered acyclic networks. One possibility to deal with networks containing cycles is to use convolutional network coding; see [23, p. 36]. Doing this, each edge has a fixed unit delay and the coding equations (4.5) get the form

$$x_i(\tau+1) = \sum_{j=1}^{l} a_{ij}x_j(\tau) + \sum_{j=1}^{r} b_{ij}u_j(\tau) \quad \text{for } i = 1,\ldots,l \tag{4.8}$$

where τ is the time variable. Since we want to consider memory-free coding, a node only receives linear combinations of the values of other nodes from the preceding time step. As now a sink node could additionally receive its own value from the preceding time step, the decoding equations are given by

$$y_{i,k}(\tau+1) = \sum_{j=1}^{l} c_{kj}^{(i)}x_j(\tau) + \tilde{c}_k^{(i)}y_{i,k}(\tau) \quad \text{for } i = 1,\ldots,d \text{ and } k = 1,\ldots,r.$$

$$\tag{4.9}$$

For $i = 1,\ldots,d$, define $\tilde{C}^{(i)} \in \mathbb{F}^{r\times r}$ as the diagonal matrix with diagonal elements $\tilde{c}_1^{(i)},\ldots,\tilde{c}_r^{(i)}$. Setting $x = (x_1,\ldots,x_l)$, $u = (u_1,\ldots,u_r)$ and $y_i = (y_{i,1},\ldots,y_{i,r})$ as well as $X_i = (x^\top \ y_i^\top)^\top$ for $i = 1,\ldots,d$, one obtains the following linear system equations

$$X_i(\tau+1) = \begin{bmatrix} A & 0 \\ C^{(i)} & \tilde{C}^{(i)} \end{bmatrix} \cdot X_i(\tau) + \begin{bmatrix} B \\ 0 \end{bmatrix} \cdot u(\tau)$$

$$y_i(\tau) = [0 \ I] \cdot X_i(\tau). \tag{4.10}$$

Defining $\mathcal{A}_i := \begin{bmatrix} A & 0 \\ C^{(i)} & \tilde{C}^{(i)} \end{bmatrix}$, $\mathcal{B} := \begin{bmatrix} B \\ 0 \end{bmatrix}$ and $\mathcal{C} := [0 \ I]$, the corresponding transfer functions are

$$G_i(z) = \mathcal{C}(zI - \mathcal{A}_i)^{-1}\mathcal{B} = (zI - \tilde{C}^{(i)})^{-1}C^{(i)}(zI - A)^{-1}B \in \mathbb{F}^{r\times r}(z).$$

According to the proof for Theorem 2.7. of [23], one has a solution for this network coding problem if and only if $\det(G_i) \not\equiv 0$ for $i = 1,\ldots,d$. Since obviously $(zI - \tilde{C}^{(i)})^{-1}$ is invertible, this is true if and only if $\det(C^{(i)}(zI - A)^{-1}B) \not\equiv 0$ for $i = 1,\ldots,d$. To achieve some necessary criteria that this condition is fulfilled, we need the following lemma, which transfers Lemma 16 to the case of convolutional network coding.

Lemma 17.
For $i = 1,\ldots,d$, the transfer function G_i of a convolutional network coding problem is not identically zero if and only if $\det\left(\begin{bmatrix} C^{(i)} & O \\ zI - A & B \end{bmatrix}\right) \not\equiv 0.$

Proof.
Analogously to the proof for Lemma 2.3 of [23], it holds

$$\left| \det \left(\begin{bmatrix} C^{(i)} & O \\ zI - A & B \end{bmatrix} \right) \right| = \left| \det \left(C^{(i)}(zI - A)^{-1}B \right) \right| \cdot \left| \det(zI - A) \right|.$$

Since $\det(zI - A)$ is monic polynomial of degree l and therefore, not identically zero, the statement follows. $\qquad\square$

Theorem 92.
For a solution of the network coding problem given by (4.8) and (4.9), it holds:

(a) *B is of full column rank,*

(b) *$C^{(i)}$ are of full row rank for $i = 1, \ldots, d$,*

(c) *(A, B) is reachable,*

(d) *$(A, C^{(i)})$ are observable for $i = 1, \ldots, d$.*

Proof.
Clearly, statements (a) and (b) are true if $\det(C^{(i)}(zI - A)^{-1}B) \not\equiv 0$ for $i = 1, \ldots, d$. Moreover, using the preceding lemma, statements (c) and (d) follow from the Hautus-test. $\qquad\square$

The preceding theorem makes it possible to achieve upper bounds for the probability to get a solution for the considered network coding problem.

Corollary 25.
The probability that random choice of the matrices $A \in \mathbb{F}^{l \times l}$, $B \in \mathbb{F}^{l \times r}$ and $C^{(i)} \in \mathbb{F}^{r \times l}$ for $i = 1, \ldots, d$ leads to a solution for the network coding problem given by (4.8) and (4.9) is upper bounded by

(a) $\left(\prod_{j=l-r+1}^{l}(1 - t^j) \right)^d \cdot \left(\prod_{j=r}^{l+r-1}(1 - t^j) \right)$

(b) $\left(\prod_{j=l-r+1}^{l}(1 - t^j) \right)^{d+1}$

Proof.
For (a) one employs that $C^{(i)}$ have to be of full row rank for $i = 1, \ldots, d$ and (A, B) has to be reachable and for (b) that B has to be of full column rank and $C^{(i)}$ have to be of full row rank for $i = 1, \ldots, d$. $\qquad\square$

Remark 36.
It depends on the values for r and l which of the preceding bounds is best, i.e. has the smallest value. Bound (b) is better than (a) if and only if $\prod_{j=l-r+1}^{l}(1 - t^j)$ is smaller than $\prod_{j=r}^{l+r-1}(1 - t^j)$. Moreover, it is easy to see that for $r \leq \frac{l+1}{2}$, bound (a) is strongest and for $r = l$, bound (b) is strongest.

Finally, we want to investigate the observability and reachability of the linear system given by (4.10).

Theorem 93.

(a) For $i = 1, \ldots, d$, $(\mathcal{A}_i, \mathcal{C})$ is observable if and only if $(A, C^{(i)})$ is observable.

(b) If $(\mathcal{A}_i, \mathcal{B})$ is reachable, then (A, B) and $(\tilde{C}^{(i)}, C^{(i)})$ are reachable. The converse, however, is not true.

Proof.

(a) The matrix $\begin{pmatrix} zI - \mathcal{A}_i \\ \mathcal{C} \end{pmatrix} = \begin{bmatrix} zI - A & 0 \\ -C^{(i)} & zI - \tilde{C}^{(i)} \\ 0 & I \end{bmatrix}$ is of full column rank if and only if $\begin{pmatrix} zI - A \\ C^{(i)} \end{pmatrix}$ is of full column rank. Thus, the statement follows from the Hautus-test.

(b) If the matrix $[zI - \mathcal{A}_i \ \mathcal{B}] = \begin{bmatrix} zI - A & 0 & B \\ -C^{(i)} & zI - \tilde{C}^{(i)} & 0 \end{bmatrix}$ is of full row rank, then $[zI - A \ B]$ and $[zI - \tilde{C}^{(i)} \ C^{(i)}]$ have to be of full row rank, too. Furthermore, a counterexample for the converse is given by $A = \begin{bmatrix} 0 & 0 \\ 1 & 0 \end{bmatrix}$, $B = \begin{bmatrix} 1 & 1 \\ 0 & 0 \end{bmatrix}$, $\tilde{C}^{(i)} = 0$ and $C^{(i)} = I_2$. $\square$

Corollary 26.

(a) For $i = 1, \ldots, d$, the probability that $(\mathcal{A}_i, \mathcal{C})$ is observable if the matrices $A \in \mathbb{F}^{l \times l}$, $B \in \mathbb{F}^{l \times r}$, $C^{(i)} \in \mathbb{F}^{r \times l}$ and $\tilde{C}^{(i)} \in \mathbb{F}^{r \times r}$ are chosen randomly is equal to $\prod_{j=r}^{l+r-1}(1 - t^j)$.

(b) For $i = 1, \ldots, d$, the probability that $(\mathcal{A}_i, \mathcal{B})$ is reachable if the matrices $A \in \mathbb{F}^{l \times l}$, $B \in \mathbb{F}^{l \times r}$, $C^{(i)} \in \mathbb{F}^{r \times l}$ and $\tilde{C}^{(i)} \in \mathbb{F}^{r \times r}$ for $i = 1, \ldots, d$ are chosen randomly is upper bounded by $(1 - t^l)^r \cdot \prod_{j=1}^{r-1}(1 - jt)\prod_{j=r}^{l+r-1}(1 - t^j)$.

Proof.

(a) This statement follows directly from Corollary 8.

(b) The probability of reachability for (A, B) is given by Theorem 26 to be equal to $\prod_{j=r}^{l+r-1}(1 - t^j)$. Since $\tilde{C}^{(i)}$ is a diagonal matrix, the reachability of $(\tilde{C}^{(i)}, C^{(i)})$ is equivalent to the reachability of the parallel connection of the scalar systems $\left(\tilde{c}_j^{(i)}, (c_{j1}^{(i)} \cdots c_{jl}^{(i)})\right)$ for $j = 1, \ldots, r$. The corresponding transfer functions are given by $(z - \tilde{c}_j^{(i)})^{-1}(c_{j1}^{(i)} \cdots c_{jl}^{(i)}) = (c_{j1}^{(i)} \cdots c_{jl}^{(i)}) \cdot diag(z - \tilde{c}_j^{(i)}, \ldots, z - \tilde{c}_j^{(i)})^{-1}$. According to Theorem 28, one needs the probability that the node systems are reachable as well as the probability of mutual left coprimeness. The node systems are reachable if and only if $(c_{j1}^{(i)} \cdots c_{jl}^{(i)}) \neq 0$ and the matrices $diag(z - \tilde{c}_j^{(i)}, \ldots, z - \tilde{c}_j^{(i)})$ are mutually left coprime if and only if the values

$\tilde{c}_j^{(i)}$ for $j = 1, \ldots, r$ are pairwise different. Consequently, the probability of reachability for $(\tilde{C}^{(i)}, C^{(i)})$ is equal to $(1 - t^l)^r \cdot \prod_{j=1}^{r-1}(1 - jt)$. Note that this probability is equal to zero if $|\mathbb{F}| = t^{-1} \leq r - 1$.

□

For a network coding problem it might be of more interest to investigate the joint reachabiliy of the systems defined by (4.10) for $i = 1, \ldots, d$, than just considering one of these systems since one wants to control all these systems with the same input. This is done in the following theorem.

Theorem 94.
For

$$\hat{A} := \begin{bmatrix} A & & 0 \\ C^{(1)} & \tilde{C}^{(1)} & \\ \vdots & & \ddots \\ C^{(d)} & 0 & \tilde{C}^{(d)} \end{bmatrix} \quad and \ \hat{B} := \begin{bmatrix} B \\ 0 \\ \vdots \\ 0 \end{bmatrix},$$

the probability that $(\hat{A}, \hat{B})$ is reachable is upper bounded by

$$(1 - t^l)^{rd} \cdot \prod_{j=1}^{rd-1}(1 - jt) \prod_{j=r}^{l+r-1}(1 - t^j).$$

Proof.
This proof could be done analogously to the preceding one. The only difference is that one has to consider the reachability of a parallel connection of rd scalar systems for the first part of the formula.

□

4.9 Conclusion

In this chapter, we considered various applications of the results of the preceding chapters in the area of convolutional codes. Furthermore, we did some new calculations dealing with such codes as well as with block codes, where we had to investigate constant matrices instead of polynomial matrices. Since the interconnection structures which are standard in the theory of linear systems are nearly the same as those used in the coding literature to concatenate convolutional codes, we could use the relationship between systems and codes to transfer probability results quite directly. If one considers linear network coding in the usual sense, the connection to linear systems is very weak. Both objects, a linear system as well as a network code, have a similarly defined transfer function but it is just a constant matrix in the case of network coding. However, if one considers convolutional network coding, one could formulate solvability criteria in terms of reachability and observability of certain linear systems, which again enables direct transfer of some results.

We chose several examples to show various applications but their number seems to be unlimited. For further research, one might consider more complicated interconnection structures such as woven convolutional codes (see [24]), look at generalizations of provided results or try to improve obtained bounds.

Bibliography

[1] Ahlswede, R.; Cai, N.; Li, R. S.-Y.; Yeung, R. W.: Network information flow, IEEE Trans. Inform. Theory **46** (2000), p. 1204-1216.

[2] Benedetto, S.; Montorsi, G.: Design of Parallel Concatenated Convolutional Codes, IEEE Trans. Communications **44** (1996), p. 591-600.

[3] Benedetto, S.; Divsalar, D.; Montorsi, G.; Pollara, F.: Serial Concatenation of Interleaved Codes: Performance Analysis, Design, and Iterative Decoding, IEEE Trans. Inform. Theory **44** (1998), p. 909-926.

[4] Benjamin, A. T.; Bennett, C. D.: The probability of relatively prime polynomials, Math. Mag. **80** (2007), p. 196-202.

[5] Callier, F. M.; Nahum, C. D.: Necessary and sufficient conditions for the complete controllability and observability of systems in series using the coprime decomposition of a rational matrix, IEEE Trans. Circuits Syst. **22** (1975), p. 90-95.

[6] Carlitz, L.: The Arithmetic of Polynomials in a Galois Field, American Journal of Mathematics **54** No. 1 (1932), p. 39-50.

[7] Climent, J.-J.; Herranz, V.; Perea, C.: A first approximation of concatenated convolutional codes from linear systems theory viewpoint, Linear Alg. Appl. **425** (2007), p. 673-699.

[8] Climent, J.-J.; Herranz, V.; Perea, C.: Linear system modelization of concatenated block and convolutional codes, Linear Alg. Appl. **429** (2008), p. 1191-1212.

[9] Climent, J.-J.; Herranz, V.; Perea, C.: Parallel concatenated convolutional codes from linear systems viewpoint, Systems & Control Letters **96** (2016), p. 15-22.

[10] Elias, P.: Coding for noisy channels, In IRE International Convention Record, pt. 4 (1955), p. 37-46.

[11] Estevan, V. T.: Complete-MDP Convolutional Codes over the Erasure Channel, PhD Thesis, Universidad de Alicante, 2010.

[12] Freudenberger, J.; Jordan, R.; Bossert, M.; Shavgulidze, S.: Serially Concatenated Convolutional Codes with Product Distance, available at http://www.researchgate.net/publication/228712396.

[13] Fuhrmann, P. A.: On controllability and observability of systems connected in parallel, IEEE Trans. Circuits and System **22** (1975), p. 57.

[14] Fuhrmann, P. A.; Helmke, U.: The Mathematics of Networks of Linear Systems, New York 2015.

[15] Garcia-Armas, M.; Ghorpade, S. R.; Ram, S.: Relatively prime polynomials and nonsingular Hankel matrices over finite fields, Journal of Combinatorial Theory, Series A **118.3** (2011), p. 819-828.

[16] Glover, K.; Silverman, L. M.: Characterization of structural controllability **AC-21** (1976), p. 534-537.

[17] Guo, X.; Yang, G.: The probability of rectangular unimodular matrices over $\mathbb{F}_q[x]$, Linear algebra and its applications **438** (2013), p. 2675-2682.

[18] Guo, X.; Hou, F.; Liu, X.: Natural density of relative coprime polynomials in $\mathbb{F}_q[x]$, Miskolc Mathematical Notes **15** No. 2 (2014), p. 481-488.

[19] Hara, S.; Hayakawa, T.; Sugata, H.: LTI systems with generalized frequency variables: A unified framework for homogeneous multi-agent dynamical systems, SICE J. of Contr. Meas. and System Integration **2** (2009), p. 299-306.

[20] Helmke, U.; Jordan, J.; Lieb, J.: Probability estimates for reachability of linear systems defined over finite fields, Advances in Mathematics of Communications **10** No. 1 (2016), p. 63-78.

[21] Helmke, U.; Jordan, J.; Lieb, J.: Reachability of random linear systems over finite fields, in Coding Theory and Applications, 4th International Castle Meeting, Palmela Castle, Portugal (eds. Pinto, R.; Malonek, P. R.; Vettori, P.), Springer-Verlag (2014), p. 217-225.

[22] Hinrichsen, D.; Prätzel-Wolters, D.: Generalized Hermite Matrices and Complete Invariants of Strict System Equivalence, SIAM J. Control Optim. **21** No. 2 (1983),p. 289-305.

[23] Ho, T; Lun, D. S.: Network Coding: An Introduction, Cambridge University Press, New York 2008.

[24] Höst, S.: Woven Convolutional Codes I: Encoder Properties, IEEE Trans. Information Theory, **48** (2002), p. 149–161.

[25] Hutchinson, R.; Rosenthal, J.; Smarandache, R.: Convolutional codes with maximum distance profile, Systems Control Lett. **54(1)** (2005), p. 53-63.

[26] Knuth, D. E.: The Art of Computer Programming, Seminumerical Algorithms, vol. 2, Addison Wesley, Reading, MA, 1969.

[27] Lidl, R.; Niederreiter, H.: Finite Fields, Cambridge University Press, Cambridge 1997.

[28] Lin, C-T.: Structural controllability, IEEE Trans. Automatic Control **19(3)** (1974), p. 201-208.

[29] Mayeda, H.: On structural controllability theorem, IEEE Trans. Automatic Control **26(3)** (1981), p. 795-798.

[30] De Reyna, J. A.; Heyman, R.: Counting tuples restricted by coprimality conditions, arXiv:1403.2769v1 (2014).

[31] Rosenbrock, H. H.: State-Space and Multivariable Theory, New York 1970.

[32] Rosenthal, J.: Connections between Linear Systems and Convolutional Codes, arXiv:math/0005281v1 (2000).

[33] Rosenthal, J.; Schumacher, J. M.; York, E. V.: On Behaviours and Convolutional Codes, IEEE Trans. Informarion Theory **42** (1996), p. 1881-1891.

[34] Rosenthal, J.; Smarandache, R.: Maximum distance separable convolutional codes, Applicable algebra in engineering, Commun. Comput. **10** (1999), p. 15-32.

[35] Rosenthal, J.; York, E. V.: BCH Convolutional Codes, IEEE Trans. Informarion Theory **45** (1999), p. 1833-1844.

[36] Schwartz, J. T.: Fast probabilistic algorithms for verification of polynomial identities, Journal of the ACM **27(4)** (1980), p. 701-717.

[37] Shields, R. W.; Pearson, J. B.: Structural controllability of multiinput linear systems, IEEE Trans. Automatic Control **AC-21** (1976), p. 203-212.

[38] Sundaram, S.; Hadjicostis, Ch.: Structural controllability and observability of linear systems over finite fields with applications to multi–agent systems, IEEE Trans. Autom. Control, **58(1)** (2013), p. 60–73.

[39] Toth, L.: The probability that k positive integers are pairwise relatively prime, Fibonacci Quart. **40** (2002), p. 13-18.

[40] Verriest, E. I.; Helmke, U: Partial state reachability of a cascade of linear systems, Preprint submitted to Automatica.

[41] Vucetic, B.; Yuan, J.: Turbo Codes, Principles and Applications, Boston 2000.

[42] Youla, D.; Pickel, P.: The Quillen-Suslin theorem and the structure of n-dimensional elementary polynomial matrices, Trans Circuits Syst **31(6)** (1984), p. 513-518.

[43] Zaballa, I.: Controllability and hermite indices of matrix pairs, Int. J. Control **68** No. 1 (1997), p. 61-68.